Plant Reproductive Ecology - Recent Advances

Edited by Anjana Rustagi and Bharti Chaudhry

Published in London, United Kingdom

IntechOpen

Supporting open minds since 2005

Plant Reproductive Ecology - Recent Advances
http://dx.doi.org/10.5772/intechopen.94800
Edited by Anjana Rustagi and Bharti Chaudhry

Contributors
Mukesh Nitharwal, Rashmi Rolania, Hanuman Singh Jatav, Kailash Chandra, Mudassar Ahmed Khan, Subhita Kumawat, Sanjay Kumar Attar, Shish Ram Dhaka, Hana'a Burezq, B.C. Anu, Ramanuj Vishwakarma, Subhankar Gurung, Arun Chettri, Sananda Mondal, Bandana Bose, Anjana Rustagi, Renu Puri, Elizabeth L. Kordyum, Helen A. Kravets

First published in London, United Kingdom, 2022 by IntechOpen
IntechOpen is the global imprint of INTECHOPEN LIMITED, registered in England and Wales, registration number: 11086078, 5 Princes Gate Court, London, SW7 2QJ, United Kingdom
Printed in Croatia

British Library Cataloguing-in-Publication Data
A catalogue record for this book is available from the British Library

Additional hard and PDF copies can be obtained from orders@intechopen.com

Plant Reproductive Ecology - Recent Advances
Edited by Anjana Rustagi and Bharti Chaudhry
p. cm.
Print ISBN 978-1-83969-493-6
Online ISBN 978-1-83969-494-3
eBook (PDF) ISBN 978-1-83969-495-0

For EU product safety concerns:
IN TECH d.o.o., Prolaz Marije Krucifikse Kozulić 3, 51000 Rijeka, Croatia, info@intechopen.com or visit our website at intechopen.com.

Meet the editors

Dr. Anjana Rustagi holds a Ph.D. in Botany from the University of Delhi, India. She has more than twelve years of experience in teaching and research. She is currently faculty in the Department of Botany, Gargi College, University of Delhi. Her research interests include regeneration and genetic transformation of crop plants for enhanced tolerance to biotic and abiotic stresses. Dr. Rustagi has also worked towards elucidating the role of ABA in embryogenesis. She has many peer-reviewed publications and book chapters to her credit. She is an advisor for Ph.D. projects and a permanent member of the International Society of Plant Morphologists.

Dr. Bharti Chaudhry is an associate professor in the Department of Botany, Ramjas College, University of Delhi, India. She was awarded a Ph.D. from the University of Delhi in 1993 and has been teaching Reproductive Biology of Angiosperms, Economic Botany, Plant Anatomy, and Plant Tissue Culture to undergraduate students in Botany and Life Sciences for more than twenty-eight years. She is the author of the books *Biology of Pollen* and *A Handbook of Common Medicinal Plants Used in Ayurveda*. She has many peer-reviewed publications and book chapters to her credit. Her current research interests are food and nutrition, herbal drugs, and Indian traditional systems of medicine.

Contents

Preface

Plant reproductive ecology is concerned with the adaptive significance of the variation in traits associated with floral biology, pollination, seed dispersal, and seedling establishment. The immobility of plants and their reliance on pollen vectors for mating is the main cause of the diversification of floral traits and pollination systems in flowering plants. This book explores the reproductive diversity of flowering plants in relation to environmental variables. It has direct relevance in agriculture, horticulture, conservation of biodiversity, control of invasive plant species, plant breeding, and biotechnology.

The volume is a compilation of seven chapters written by eminent scientists across the world. Flower symmetry is a key evolutionary innovation of angiosperms and transition from radial to bilateral symmetry is believed to be associated with increased specificity to pollinators, thereby favoring reproductive isolation. Chapter 1 unravels the genetic, phylogenetic, and evolutionary patterns for transition in flower symmetry. Chapter 2 provides an interesting and comprehensive review of the evolutionary transformations of the reproductive structures of angiosperms in light of aromorphosis, allomorphosis, specialization, and reduction, which are the key directions of the evolutionary process.

Pollination is an integral event for the reproductive success of a plant. Chapter 3 focuses on the diversity of pollinators in an agro-ecosystem, the interactions between plants and pollinators, and the threat of declining pollinators. The authors highlight the need for comprehensive management strategies and restoration of pollinators for sustainable agriculture. Chapter 4 reviews the complex breeding system of commercially important Citrus species like mandarin, sweet orange, pummelo, grapefruit, lemons, limes, and citrons and their dependency on pollination services in a global pollinator decline scenario. Chapter 5 discusses the complex interactions and impact of non-native exotic plant communities on native plants and pollinators.

Seed biology is an important aspect of reproductive ecology, as fruits and seeds are the economic products of many of our crop plants. Chapter 6 highlights recent advances in the reproductive ecology of *Medicago sativa* (alfalfa), an important forage legume, in relation to its floral biology, pollination ecology, ovule sterility, inbreeding depression, breeding strategies, seed production, and seed dispersal. Chapter 7 focuses on seed priming, a cost-effective interlinking technology between seed germination and seedling establishment. The authors discuss the relevance of the technique in enhancing seed vigour and seed germination and promoting successful crop stand establishment under extreme climatic conditions and fragile ecosystems.

The book is a useful resource for ecologists, botanists, zoologists, plant breeders, and conservation biologists.

We thank Author Service Manager, Ms. Jasna Bozic, at IntechOpen for her constant support and help in completing this project.

Anjana Rustagi
Department of Botany,
Gargi College,
University of Delhi,
Delhi, India

Bharti Chaudhry
Department of Botany,
Ramjas College,
University of Delhi,
Delhi, India

Chapter 1

Current Trends in Developmental Genetics and Phylogenetic Patterns of Flower Symmetry

Renu Puri and Anjana Rustagi

Abstract

Flower Symmetry is a key evolutionary innovation in some lineages of angiosperms. The flowers of the primitive angiosperm plants were radially symmetrical actinomorphic. Later bilaterally symmetrical zygomorphic flowers independently evolved in several clades of angiosperms. This transition of trait is associated with an adaptation to specialized methods of pollination. Zygomorphic flowers allow more specific plant insect interaction. So, the transition from radial symmetry to bilateral symmetry facilitates reproductive isolation which in turn might have led to diversification or rapid speciation of some lineages in angiosperms. Phylogenetic analyses in lineages of angiosperms revealed that few clades have shown that there have been reversals, that is, there is transition from bilateral symmetry to radial symmetry. When such studies are correlated with genetic studies, it is revealed that CYC (TCP family) transcription factors are responsible for the transition of this floral trait. Phylogenetic analyses, genetic studies and Evo-Devo analyses can answer important questions such as what other transition in floral symmetry is found in angiosperms? Is there a pattern of floral symmetry transition in different lineages? Do these transitions act as key innovation for the clades in which they have evolved?

Keywords: floral symmetry, zygomorphic, actinomorphic, phylogeny, asymmetry

1. Introduction

Flower is a significant novelty for evolutionary success in angiosperms. It primarily comprises four whorls—sepals, petals, stamens, and carpels. The shape of the flower changes because of change in the shape or morphology of any of these whorls. This gives rise to different shape and symmetry of flowers. The change in symmetry can occur in any of the whorl; however, it is widely studied in petals [1]. Floral symmetry is an important trait as it impacts the visual appearance of a flower. Hence, it's been a fascination for human eye. Pollinators are usually attracted to flowers due to its diverse forms of colors but also due to the symmetry it possesses, thereby contributing to the plant pollination syndrome [2–4]. Broadly there are the two types of floral symmetry, radial symmetry also known as polysymmetry or actinomorphy and bilateral symmetry also known as monosymmetry or zygomorphy. Flowers with radial symmetry have more than two planes of symmetry and are called as actinomorphic. Flowers with bilateral symmetry have single plane of symmetry and are called as zygomorphic [5–7]. There is another rare form of symmetry

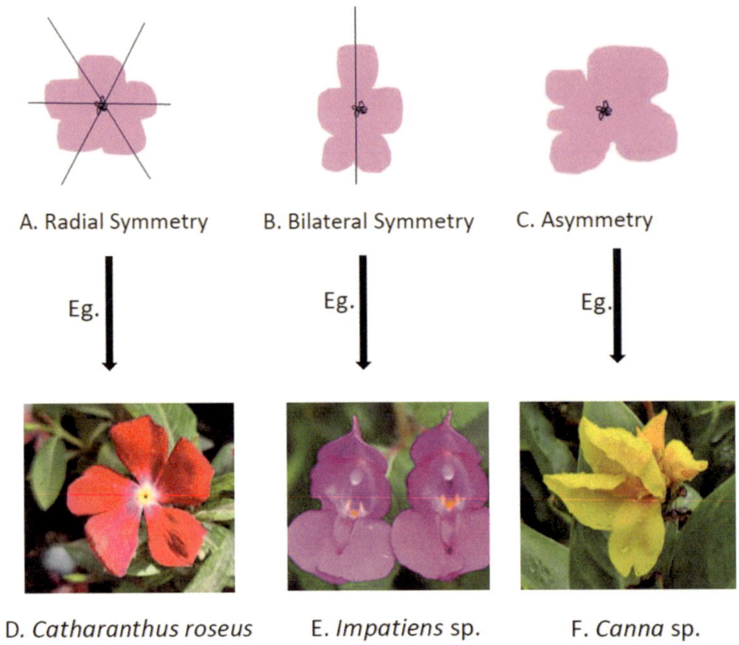

Figure 1.
Different types of floral symmetry. A. Radial symmetry (more than one plane of symmetry, polysymmetry, actinomorphy). B. Bilateral symmetry (single plane of symmetry, monosymmetry, zygomorphy), C. asymmetry (no plane of symmetry). D. Flower of Catharanthus roseus *showing radial symmetry, E. flower of* impatiens *sp. showing bilateral symmetry and F. flower of* canna *sp. showing asymmetry.*

in flowers that is known as asymmetry [8]. This refers to morphologies where there is no pane of symmetry (**Figure 1**).

These categories have been studied at various different levels such as the molecular aspects of these transitions and how the pollinators perceive them [5, 6]. There have been transitions from actinomorphy to zygomorphy many times during the diversification of angiosperms, and these transitions are more common in species-rich lineages such as Fabaceae, Lamiales, and Orchidaceae. However, reversals from zygomorphy to actinomorphy are also reported [9–12].

We in this chapter focus on the different genetic studies, which have been conducted to understand the molecular basis of the variation in floral symmetry and what do we get to know when these studies are correlated with phylogenetic studies. These studies have provided insights into how and when these transitions in floral symmetry evolve.

2. Diversity of floral symmetry in angiosperm flowers

Apart from the flower symmetry categories mentioned above, there are many other forms of flower symmetry such partial zygomorphy and few others. There are different degrees of symmetry. In the year 1925, based on the aspects of symmetry as used in crystallography, new terms were introduced. Of those rotational symmetry, mirror symmetry and spiral symmetry are to name a few [13].

Later correlation studies between floral symmetry and pollination biology were conducted. These studies focused on how pollinators perceived flowers. With these studies, three-dimensional aspects were added to the floral symmetry terminology [10, 14, 15]. After around 75 years, an elaborated and modified classification was proposed, which was also based on visual perception of flowers by the pollinators [2, 3]. These terminally only applied to very discreet flower forms and only in mature flowers [16].

However, there are variable degrees of floral symmetry at different developmental stages. This variation can also be seen in different lineages, or there might be convergent evolution of this state in two different and closely unrelated clades. Endress [5] considered these two aspects, that is, developmental changes and phylogenetic changes, and identified three forms of monosymmetry and three forms of asymmetry.

First form of monosymmetry is found in taxa with elaborated monosymmetric flowers, for example, Lamiales, Asterales, and Leguminosae. The second is taxa in which monosymmetry arises, but predominantly the group is polysymmetric, e.g., *Passiflora lobata* (Passifloraceae) and *Chiranthodendron pentadactylon* (Malvaceae). The third form is evolution of monosymmetry form by reduction, e.g., in case of *Hippuris* (Antirrhinaceae). First form of asymmetry is seen in taxa, which are predominantly monosymmetric, e.g., *Vigna* (Leguminosae). Second form is unordered asymmetry in flowers of basal angiosperms, e.g., *Zygogynum* (Winteraceae). Third form is asymmetry, which arises due to reduction, e.g., *Centranthus* (Valerianaceae) [5].

3. The genetic basis of flower symmetry

The genetics of a flower is regulated by specific transcription factors (TFs) [17]. TFs such as MADS (First alphabet of MCM1 in yeast, AGAMOUS in Arabidopsis, DEFICIENS in snapdragon, and SERUM RESPONSE FACTOR in human)-box are widely studied for various developmental pathways from root development to fruit development. One important role is determination of organ development in flower. ABCDE model and its modifications are based on the different functions of *MADS*-box TFs [18–25].

Floral development is also controlled by other set of TFs Known as *MYB* TFs. These TFs are found in all eukaryotic organisms and identified by the presence of *MYB* R Repeats. Each repeat is about 52 amino acids. Based on these repeats, the *MYB* TFs are classified into 4R, 3R, 2R, and 1R-MYB types. In plants, the most common are 2R-*MYB* TFs (R2R3) [26–29].

Recent studies show that *MYB* TFs, *DIV*-and-RAD-interacting-factors (*DRIF)*, *DIVARICATA (DIV)* and, *RADIALIS (RAD)* play important role in floral symmetry [30, 31]. The studies revealed that these MYB TFs interact with TCP (TCP family name is derived from its first three characterized members, first alphabet of TB1, TEOSINTE BRANCHED 1; CYC, CYCLOIDEA and PCFs, proliferating cell factors) TFs. *CYCLOIDEA (CYC)* and *DICHOTOMA (DICH)* are two paralogs that belong to TCP family. They are expressed in the dorsal region of flowers, and they are vital to control floral symmetry [32]. The studies show that these TFs and their orthologs and paralogs have similar interactions in Dicots and monocots.

3.1 Developmental genetics of floral symmetry in dicots

The molecular basis of floral symmetry was first studied in *Antirrhinum majus,* which has zygomorphic flowers [33]. In the dorsal region, there is interaction of *RAD* and *DRIF*. *CYC* is expressed only in the dorsal region of flower. *RAD* promoter and intron have *CYC* target sequence. When *CYC* binds to *RAD* promoter, its causes synthesis of RAD protein. The DIV protein Interacts with *DRIF* but when *RAD* binds to *DRIF*, *DIV/DRIF* complex is not formed and thus not able to activate downstream ventral gene. In the ventral region, *CYC* is not expressed and likewise *RAD* is not activated. *DRIF* is free to bind to its target sequence present on the *DIV* promoter region [34, 35]. This heterodimer complex, *DIV/DRIFT* complex activates the ventral genes. This differential expression of *CYC* in dorsal and ventral region of snapdragon flower causes dorsoventral symmetry (**Figure 2**) [36–38].

Flower of *A. majus* is an example from Lamiales where there is elaborated zygomorphy. Within Lamiaceae in family Gesneriaceae, the clade has zygomorphic flowers. However, *Conandron ramondioides* have actinomorphic flowers [39]. This is the case of reversal from zygomorphy to actinomorphy. In this species, there is change in the expression of the homologs of above TFs. In case of petals and stamen, there is loss of expression of *CrCYC* and *CrRAD* so the *DIV* is active and ventral genes are activated (**Figure 3**) [40–42].

Another member of clade lamiales *Plantago lanceolata* shows actinomorphic flowers [43]. Interestingly here *CYC* A clade gene is absent, but *PlCYC*-B clade gene is expressed. *PlCYC*-B at early stages is present in ground tissue, and later it expresses in stamens till the upper portion of filament. *PlCYC* is absent in petals. Homolog of *RAD* gene is absent, and *PlDIV*, ortholog of *DIV* is expressed in lateral petals and in the ovary. Absence of *CYC-A* clade and *RAD* gene in the dorsal part of flower might have been responsible for radial symmetry (**Figure 4**) [44, 45].

More complex mechanism takes place in Asteraceae. Here the inflorescence is complex and is known as capitulum. For example, in *Senecio vulgaris,* the disc florets have radial symmetry and ray florets have bilateral symmetry [46]. Here there are CYC-like genes *RAY1, RAY2,* and *Ray3*. Unlike as in *A. majus, RAY1* and *RAY2* are expressed in entire ray floret, and RAY3 expresses only in ventral region. MYB genes *SvDIV1B* and *SvRAD* are expressed only in the ray florets at early stages. *SvRAD* expresses in ventral region. At later stages, *SvDIV1B* expresses in disc florets too (**Figure 5**) [47].

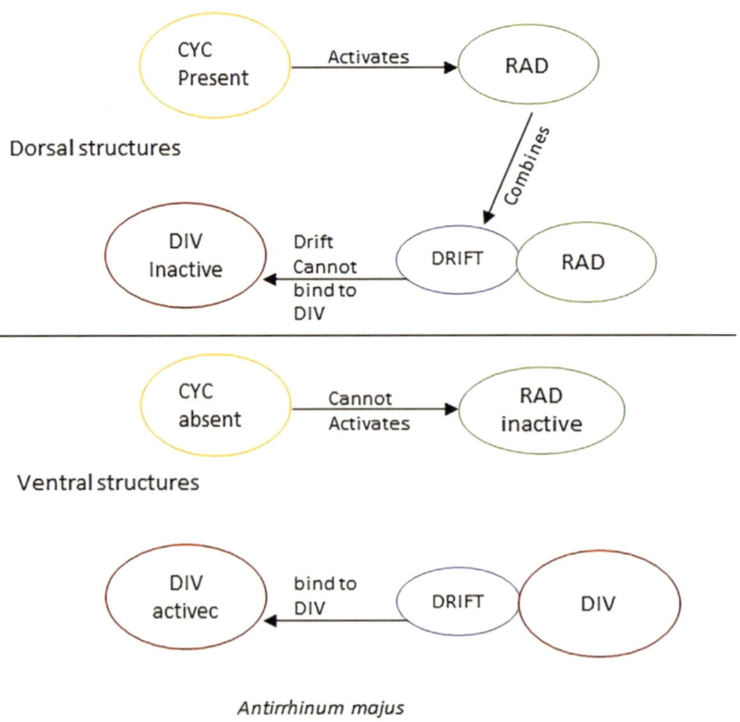

Figure 2.
Molecular genetic control of floral symmetry in Antirrhinum majus.

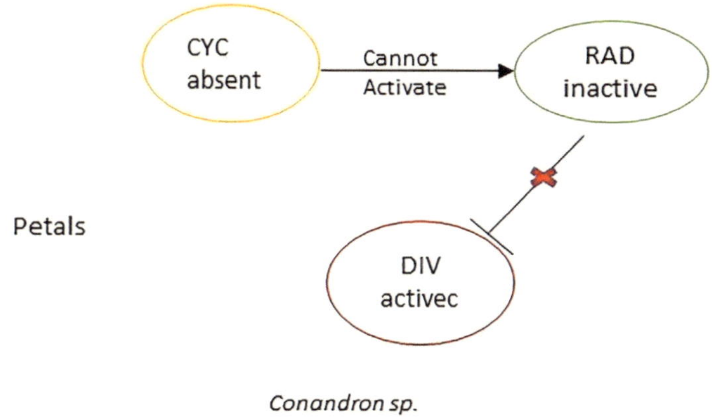

Figure 3.
Molecular genetic control of floral symmetry in Conandron *sp.*

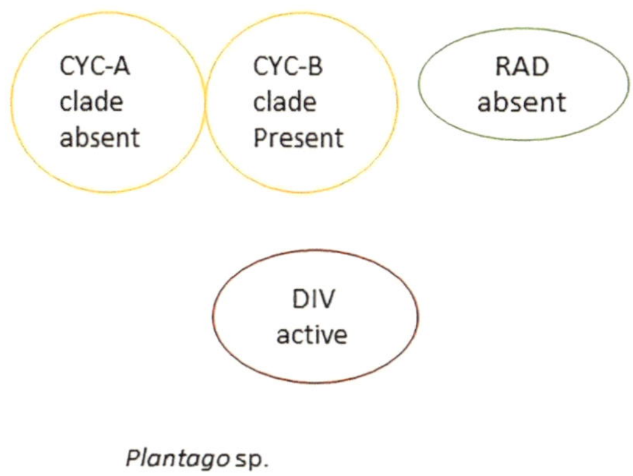

Figure 4.
Molecular genetic control of floral symmetry in Plantago *sp.*

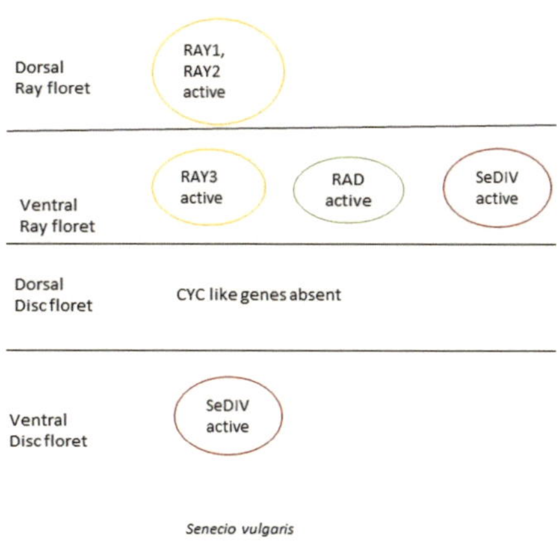

Figure 5.
Molecular genetic control of floral symmetry in Senecio vulgaris.

3.2 Developmental genetics of floral symmetry in monocots

Little is known about floral symmetry in monocots. Orchidaceae family supports the DDR (DDR stands for DIV, RAD, and DIV-and-RAD-Interacting Factor DRIF)

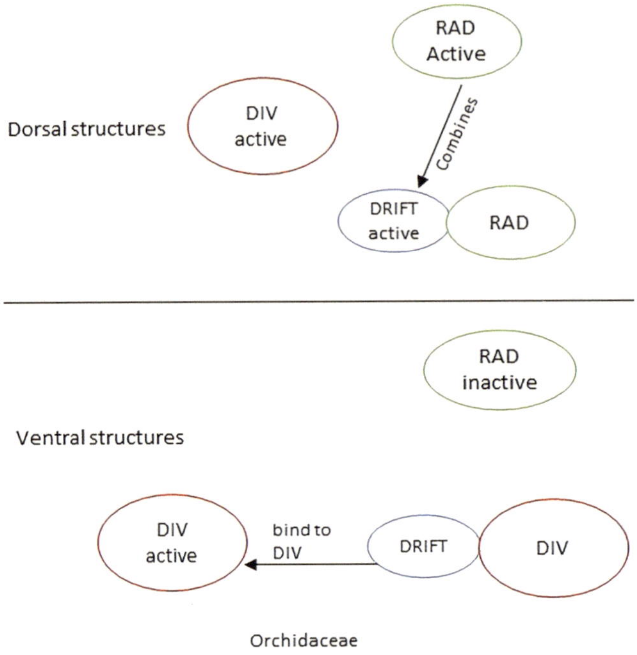

Figure 6.
Molecular genetic control of floral symmetry in Orchidaceae.

regulatory module [48]. Recent study revealed that *DIV, DRIFT,* and *RAD* interact with each other and work as regulatory unit. DDR module is more conserved in monocots than in dicots. Orchids show resupination before anthesis and dorsally placed structure becomes ventral in position. *DIV* and *DRIF* express in ventral structures and after resupination take dorsal position. In the lip high level of *RAD* expression prevents activation of *DIV.* RAD is responsible for lip determination in orchids (**Figure 6**) [49].

4. Other putative genes

Apart from *MYB* and *TCP* family, other putative genes need to be identified. Few genes are identified in model plant Arabidopsis. One such is *RABBIT EARS* (*RBE*). It belongs to C2H2 Zinc finger TFs family. Its expression is regulated by Auxin. When this gene is muted, then two of the petals do not elongate and give rise to bilateral flower. *RBE* further regulates *TCP4* negatively by binding directly to its target sequence to the *TCP4* promoter [41]. The role of CYC and *CYC*-like genes, which belong to *TCP* family, is already known. Other putative *TCP* members could be analyzed to reveal other pathways responsible for flower symmetry.

Other putative genes are such as *AINTEGUMENTA-LIKE 6* (*AIL6*), *AUXIN-REGULATED GENE INVOLVED IN ORGAN GROWTH* (*ARGOS*), *AINTEGUMENTA* (*ANT*), *BIGPETALp* (*BPEp*), *and JAGGED* (*JAG*). The differential expression of these genes plays important role in floral organ development. Further analyses on non-model species that have different floral symmetries can reveal their role in floral symmetry [41]. The identification of these genes will not only help to identify the QTL for floral traits but also for phylogenetic studies. The homologs of these genes can further help us to understand the evolution of these genes and the gene families.

5. Phylogenetic pattern in floral symmetry

Floral symmetry patterns are best understood in phylogeny context. These studies help in understanding how often the transition from radial symmetry to bilateral symmetry has occurred and vice versa. It also gives insight on what lineages these transitions have taken place and when those transitions occurred on geological timescale.

We now have clear understanding about the major lineages of angiosperms. Recent studies focus on mapping various morphological traits on these robust phylogenetic analyses. In relation to floral symmetry, recent studies have constructed it as a character and different forms (radial symmetry, bilateral symmetry, asymmetry, etc.) on phylogenies [50]. Such robust studies have answered the abovementioned questions.

Studies revealed that the ancestral flower of angiosperms was radially symmetrical [8]. Floral symmetry character reconstruction on ordinal phylogeny also revealed the same scenario and showed that the transition to bilateral symmetry is widespread on angiosperm phylogeny. Parsimony reconstruction on family phylogeny revealed that there are at least 70 such transitions from radial to bilateral symmetry in angiosperms including 23 in monocots and 46 in eudicots [50].

Later studies focused on detailed phylogeny of smaller clades. Character reconstruction of floral symmetry in Lamiales at family level revealed one transition from radial to bilateral symmetry and one vice versa [51]. Multiple transitions from radial to bilateral symmetry were observed in Brassicaceae, Ranunculaceae, and Solanaceae [52–54].

S.no	Clade names
Group 1	Amborrellales, Pandanales, Arecales, Dasypogonaceae, ceratophyllales, Trochodendrales, Buxales, Gunnerales, Dileniales, Vitales, Celastrales, Oxalidales, Fagales, Crossosomatales Picramniales, Heurteales, Malvales, Berberidopssidales, Cornales, Nymphaellales, Austrabaileyales, Chaloranthales, Canellales, Magnoliales, Petrosaviales, Aquifoliales, Escalloniales, Bruniales, Paracryphiales, Garryales, Icacinaceae, Metteniusaceae, Oncothecaceae, Vahliaceae

Table 1.
List of clades in which radial symmetry is conserved in all or most of its descendents.

S.no	Clade names
Group 2	Piperales, Laurales, Acorales, Alismatales, Dioscoreales, Liliales, Asparagales, Poales, Commelinales, ranunculales, Proteales, Sabiales, Saxifragales, Zygophyllales, Malphighiales, Fabales, Rosales, Cucurbitales, Geraniales, Myrtales, Sapindales, Brassicales, Santalales, Caryophyllales, Ericales. Apiales, Dipsacales, Asterals, Boraginaceae, Gentianales, Solanales

Table 2.
List of clades in which bilateral symmetry evolved independently.

S.no	Clade names
Group 3	Zingiberales, Lalmiales Acorales, Fabales, Dispacales

Table 3.
List of clades in which bilateral symmetry evolved early as a single event.

S.no	Clade names
Group 2	Orchidaceae, Fabaceae, Malpighiaceae,Capparaceae, Cleomaceae, Brassicaceae, Caprifoliaceae and Aster-aceae

Table 4.
List of clades in which there was reversal from bilateral symmetry to radial symmetry.

Bilateral symmetry has evolved at least 130 times independently in different clades, and there were at least 70 reversals [55]. Based upon these transitions, four basic groups have been observed. These are: first, there are clades where radial symmetry is conserved (**Table 1**). Second, clades wherein bilateral symmetry has evolved independently (**Table 2**). Third, clades wherein bilateral symmetry arises as single early event (**Table 3**), and fourth group includes clades that show reversal to radial symmetry (**Table 4**). Basal angiosperms have radial symmetry with exceptions such as in *Glossocalyx.* Radial symmetry is also conserved in ancestral monocots and Eudicots. More robust phylogenetic studies in future can further reveal detailed transitions.

6. Floral symmetry on geological timescale

First fossil remains of flowers with inserted flower parts are found to be from early Cretaceous period (Barremian-Aptian period) around 125 million years ago (Ma.). This fossil represents the flowers of the ancestors of early Nympheales [56, 57]. The first fossil of a flower, which is pentamerous, was reported from late Cretaceous period (Cenomanian period) around 100 Ma. This fossil remains has both petals and sepals. It is considered as a representative of ancient ancestor of Eudicots [58].

Clearly, the above fossil records show that there was a transition from closed floral structure to an open floral structure. The flower evolved from closed noncyclic structures to more open and cyclic forms. This transition took place in Mid-Cretaceous period. It is during this period that many floral traits evolve. Many of these traits are key innovations. This floral trait evolution coincides with the major diversification period of angiosperms [59, 60].

The first transition from radial symmetry to bilateral symmetry can be traced to the first radiation in angiosperms. These flower remains are reported from Turonian fossils from late Cretaceous, which is around 100 Ma. These flower fossils have staminodal nectaries making the radial flower partially bilateral. These flowers fossils are the first report of zygomorphic flower form, although these forms were not exactly bilaterally symmetrical. These fossils represented the precursors or ancestors of zygomorphic flowers [59, 61].

First complete zygomorphic flowers are recorded from Paleogene (Paleocene-Eocene period) around 55 Ma [59]. This is the time period where a second major diversification of angiosperms took place. So, we have seen that both the radiation events of angiosperms diversification coincide with the evolution of floral traits including floral symmetry [61]. Thus, floral symmetry transition is clearly the key innovation, which might have played crucial role in radiation of angiosperms. Of course, there might be other factors too that played their part. One such factor is evolution of pollinators.

Interestingly, the evolution of those floral traits that lead to diversification of angiosperms in the above-said events coincides with the advent of specialized pollinators. Also, the evolution of bilateral symmetry in some plant lineages cooccurs with the time period when there was a rise of some bee families. Thus, in some lineages the Coevolution of insect pollinators with the floral symmetry holds true [2, 5]. Although there are other abiotic and biotic factors that are needed to be taking into account such as climatic conditions and various architectural components of flower.

7. Conclusion

Great deal of progress is being made on study of floral symmetry evolution in dicots. The *MYB* TFs and *TCP*s play an important role in understanding the molecular genetic basis of floral symmetry. However, there is a scope to identify other putative genes that might be having evolutionary significance. Robust studies should be taken up, especially among monocots to unravel modulators of floral symmetry patterns.

Most of the species-rich lineages have bilateral symmetry. Based on this observation, many hypothesize that bilateral symmetry is related to increased specificity to pollinators, thereby increasing the chances of reproductive isolation. This holds true for many taxa, but not all lineages of angiosperms follow the same pattern.

Evolution is a highly complex phenomenon, and diversification of species is dependent on various factors and not just one single trait. Therefore, it is necessary to take holistic approach and to combine other factors such as developmental stages, floral mechanics, etc., with the phylogenetic framework to get a detailed answers about floral symmetry.

Acknowledgements

The authors acknowledge the DST Project Grant (file no. ECR/2017/000563) for financial support during conceptualization and manuscript preparation.

Author details

Renu Puri and Anjana Rustagi*
Department of Botany, Gargi College, University of Delhi, Delhi, India

*Address all correspondence to: anjana.rustagi@gargi.du.ac.in

References

[1] Moyroud E, Glover BJ. The evolution of diverse floral morphologies. Current Biology. 2017;**27**:R941-R951. DOI: 10.1016/j.cub.2017.06.053

[2] Neal PR, Dafni A, Giurfa M. Floral symmetry and its role in plant-pollinator systems: Terminology, distribution, and hypotheses. Annual Review of Ecology and Systematics. 1998;**29**:345-373

[3] Giurfa M, Dafni A, Neal PR. Floral symmetry and its role in plant-pollinator systems. International Journal of Plant Sciences. 1999;**160**:S41-S50

[4] Kalisz S, Ree RH. Sargent RD Linking floral symmetry genes to breeding system evolution. Trends in Plant Science. 2006;**11**:568-573

[5] Endress PK. Symmetry in flowers: Diversity and evolution. International Journal of Plant Sciences. 1999; **160**:S3-S23

[6] Endress PK. Evolution of floral symmetry. Current Opinion in Plant Biology. 2001;**4**:86-91

[7] Endress PK. Angiosperm floral evolution: Morphological developmental framework. In: Advances in Botanical Research. Vol. 44. Amsterdam, The Netherlands: Elsevier; 2006. pp. 1-61

[8] Endress PK. The immense diversity of floral monosymmetry and asymmetry across angiosperms. The Botanical Review. 2012;**78**:345-397

[9] Stebbins GL. Flowering Plants: Evolution above the, Species Level. Cambridge, MA: Harvard University Press; 1974

[10] Leppik EE. Origin and evolution of bilateral symmetry in flowers. In: Dobzhansky T, Hecht MK, Steere WC, editors. Evolutionary Biology. New York, NY: Appleton-Century-Crofts; 1972

[11] Delpino F. Zigomorfia florale e sue cause. Malpighia. 1887;**1**:245-262

[12] Church AH. Types of Floral Mechanism. Oxford, UK: Clarendon; 1908

[13] Frey-Wyssling A. Geometrische Symmetriebetrachtung. Flora. 1925;**120**:87-98

[14] Leppik EE. Some viewpoints on the origin and evolution of flowering plants. Acta Biotheoretica. 1955;**9**:45-56

[15] Leppik EE. Evolutionary relationships between entomophilous plants and anthophilous insects. Evolution. 1957;**11**:466-481

[16] Endress ME, Albert VA. Amorphological cladistic study of Apocynaceae: Trends in character evolution within a broadened familial circumscription. American Journal of Botany. 1995;**82**(suppl):127

[17] Jin J, Tian F, Yang DC, Meng YQ, Kong L, Luo J, et al. PlantTFDB 4.0: Toward a central hub for transcription factors and regulatory interactions in plants. Nucleic Acids Research. 2017;**45**:D1040-D1045

[18] Aceto S, Gaudio L. The MADS and the beauty: Genes involved in the development of orchid flowers. Current Genomics. 2011;**12**:342-356

[19] Bowman JL, Smyth DR, Meyerowitz EM. Genetic interactions among floral homeotic genes of arabidopsis. Development. 1991; **112**:1-20

[20] Mondragon-Palomino M, Theissen G. MADS about the evolution of orchid flowers. Trends in Plant Science. 2008;**13**:51-59

[21] Riechmann JL, Meyerowitz EM. MADS domain proteins in plant development. Biological Chemistry. 1997;**378**:1079-1101

[22] Chanderbali AS, Berger BA, Howarth DG, Soltis PS, Soltis DE. Evolving ideas on the origin and evolution of flowers: New perspectives in the genomic era. Genetics. 2016; **202**:1255-1265

[23] Valoroso MC, Censullo MC, Aceto S. The MADS-box genes expressed in the inflorescence of Orchis italica (Orchidaceae). PLoS One. 2019; **14**:e0213185

[24] Gonda TJ, Bishop JM. Structure and transcription of the cellular homolog (c-MYB) of the avian myeloblastosis virus transforming gene (v-MYB). Journal of Virology. 1983;**46**:212-220

[25] Gonda TJ, Gough NM, Dunn AR, de Blaquiere J. Nucleotide sequence of cDNA clones of the murine MYB proto-oncogene. The EMBO Journal. 1985;**4**:2003-2008

[26] Dubos C, Stracke R, Grotewold E, Weisshaar B, Martin C, Lepiniec L. MYB transcription factors in Arabidopsis. Trends in Plant Science. 2010;**15**:573-581

[27] Jiang C, Gu J, Chopra S, Gu X, Peterson T. Ordered origin of the typical two- and three-repeat MYB genes. Gene. 2004;**326**:13-22

[28] Lipsick JS. One billion years of MYB. Oncogene. 1996;**13**:223-235

[29] Rosinski JA, Atchley WR. Molecular evolution of the MYB family of transcription factors: Evidence for polyphyletic origin. Journal of Molecular Evolution. 1998;**46**:74-83

[30] Raimundo J, Sobral R, Bailey P, Azevedo H, Galego L, Almeida J, et al. A subcellular tug of war involving three MYB-like proteins underlies a molecular antagonism in Antirrhinum flower asymmetry. Plant Journal: For Cell and Molecular Biology. 2013;**75**:527-538

[31] Raimundo J, Sobral R, Laranjeira S, Costa MMR. Successive domain rearrangements underlie the evolution of a regulatory module controlled by a small interfering peptide. Molecular Biology and Evolution. 2018; **35**:2873-2885

[32] Corley SB, Carpenter R, Copsey L, Coen E. Floral asymmetry involves an interplay between TCP and MYB transcription factors in Antirrhinum. PNAS. 2005;**102**(14):5068-7335

[33] Carpenter R, Coen ES. Floral homeotic mutations produced by transposon-mutagenesis in Antirrhinum majus. Genes & Development. 1990;**4**:1483-1493

[34] Eguen T, Straub D, Grae M, Wenkel S. MicroProteins: Small size-big impact. Trends in Plant Science. 2015;**20**:477-482

[35] Staudt AC, Wenkel S. Regulation of protein function by 'microProteins'. EMBO Reports. 2011;**12**:35-42

[36] Luo D, Carpenter R, Copsey L, Vincent C, Clark J, Coen E. Control of organ asymmetry in flowers of Antirrhinum. Cell. 1999;**99**:367-376

[37] Busch A, Zachgo S. Flower symmetry evolution: Towards understanding the abominable mystery of angiosperm radiation. Bioessays. 2009;**31**:1181-1190

[38] Costa MM, Fox S, Hanna AI, Baxter C, Coen E. Evolution of regulatory interactions controlling floral asymmetry. Development. 2005;**132**:5093-5101

[39] Hileman LC. Trends in flower symmetry evolution revealed through

phylogenetic and developmental genetic advances. Philosophical Transactions of the Royal Society B: Biological Sciences. 2014;**369**:20130348

[40] Galego L, Almeida J. Role of DIVARICATA in the control of dorsoventral asymmetry in Antirrhinum flowers. Genes & Development. 2002;**16**:880-891

[41] Spencer V, Kim M. Recycling molecular regulators in the evolution and development of flower symmetry. Seminars in Cell & Developmental Biology. 2018;**79**:16-26

[42] Hsin KT, Wang CN. Expression shifts of floral symmetry genes correlate to flower actinomorphy in East Asia endemic Conandron ramondioides (Gesneriaceae). Botanical Studies. 2018;**59**:24

[43] Reardon W, Gallagher P, Nolan KM, Wright H, Cardenosa-Rubio MC, Bragalini C, et al. Diverent outcomes for the MYB floral symmetry genes DIVARICATA and RADIALIS during the evolution of derived actinomorphy in Plantago. The New Phytologist. 2014;**202**:716-725

[44] Preston JC, Martinez CC, Hileman LC. Gradual disintegration of the floral symmetry gene network is implicated in the evolution of a wind-pollination syndrome. Proceedings of the National Academy of Sciences of the United States of America. 2011;**108**:2343-2348

[45] Reardon W, Fitzpatrick DA, Fares MA, Nugent JM. Evolution of flower shape in Plantago lanceolata. Plant Molecular Biology. 2009;**71**:241-250

[46] Trow AH. On the inheritance of certain characters in the common groundsel—Senecio vulgaris, Linn.—And its segregates. Journal of Genetics. 1912;**2**:239-276

[47] Garces HM, Spencer VM, Kim M. Control of floret symmetry by RAY3, SvDIV1B, and SvRAD in the capitulum of Senecio vulgaris. Plant Physiology. 2016;**171**:2055-2068

[48] Madrigal Y, Alzate JF, Gonzalez F, Pabon-Mora N. Evolution of RADIALIS and DIVARICATA gene lineages in flowering plants with an expanded sampling in non-core eudicots. American Journal of Botany. 2019;**106**:334-351

[49] Mondragon-Palomino M, Theissen G. Why are orchid flowers so diverse? Reduction of evolutionary constraints by paralogues of class B floral homeotic genes. Annals of Botany. 2009;**104**:583-594

[50] Citerne H, Jabbour F, Nadot S, Damerval C. The evolution of floral symmetry. In: Kader JC, Delseny M, editors. Advances in Botanical Research. Vol. 54. London, UK: Academic Press Ltd–Elsevier Science Ltd; 2010. pp. 85-137

[51] Schaferhoff B, Fleischmann A, Fischer E, Albach DC, Borsch T, Heubl G, et al. Towards resolving Lamiales relationships: Insights from rapidly evolving chloroplast sequences. BMC Evolutionary Biology. 2010;**10**:352. DOI: 10.1186/1471-2148-10-352

[52] Busch A, Horn S, Muhlhausen A, Mummenhoff K, Zachgo S. Corolla monosymmetry: Evolution of a morphological novelty in the Brassicaceae family. Molecular Biology and Evolution. 2012;**29**:1241-1254. DOI: 10.1093/molbev/msr297

[53] Damerval C, Nadot S. Evolution of perianth and stamen characteristics with respect to floral symmetry in Ranunculales. Annals of Botany. 2007;**100**:631-640. DOI: 10.1093/aob/mcm041

[54] Olmstead RG, Bohs L, Migid HA, Santiago-Valentin E, Garcia VF, Collier SM. A molecular phylogeny of

the Solanaceae. Taxon. 2008; **57**:1159-1181

[55] Reyes E, Sauquet H, Nadot S. Perianth symmetry changed at least 199 times in angiosperm evolution. Taxon. 2016;**65**:945-964

[56] Crepet WL. Timing in the evolution of derived floral characters: Upper Cretaceous (Turonian) taxa with tricolpate and tricolpate-derived pollen. Review of Palaeobotany and Palynology. 1996;**90**:339-359

[57] Crepet WL. The fossil record of angiosperms: Requiem or renaissance? Annals of the Missouri Botanical Garden. 2008;**95**:3-33

[58] Basinger JF, Dilcher DL. Ancient bisexual flowers. Science. 1984; **224**:511-513

[59] Crepet WL, Niklas KJ. Darwin's second "abominable mystery": Why are there so many angiosperm species? American Journal of Botany. 2009;**96**: 366-381

[60] Friis EM, Pedersen KR, Crane PR. Diversity in obscurity: Fossil flowers and the early history of angiosperms. Philosophical Transactions of the Royal Society B: Biological Sciences. 2010; **365**:369-382

[61] Dilcher D. Toward a new synthesis: Major evolutionary trends in the angiosperm fossil record. Proceedings. National Academy of Sciences. United States of America. 2000;**97**:7030-7036

Chapter 2

Evolutionary Patterns of the Internal Structures of Generative Organs in Angiosperm Plants

Elizabeth L. Kordyum and Helen A. Kravets

Abstract

Evolutionary patterns of the internal structures of generative organs in angiosperms are considered in light of the idea on their close dependence on the appearance of angiospermy – formation of the ovary closed cavity by carpels–macrosporophylls. A characteristic feature of the sexual process in gymno- and angiosperms is the independency of water for fertilization, unlike all lower plants and pteridophytes. The main direction of the further evolution of the sexual process consisted in the modification for adaptations that ensure the successful fertilization in new conditions. The guidelines and levels of evolution include aromorphosis, allomorphosis, specialization and reduction which are considered to be concrete examples of microstructure of generative organs.

Keywords: embryology, fertilization, aromorphosis, allomorphosis, specialization and reduction, angiosperms

1. Introduction

The current state of botanical knowledge allows us to believe the opinion that the appearance of angiosperms was the largest arogenesis and followed many branches with a pronounced heterochrony of the evolution of their organs [1, 2]. One of the fundamental changes in the structure of the reproductive organs of angiosperms was the formation of a closed ovary by carpels-macrosporophyles, in which the ovules were formed. The presence of a "wet chamber" protected the ovules and embryos during the seed formation was a decisive condition for the rapid dispersal of angiosperms under increasing dryness of the atmosphere at certain geological periods of the Earth [3].

Differentiation of pistil-forming carpels into closed ovary, style, and stigma also caused a change in the conditions for pollen germination and pollen tube growth, i.e. conditions for the delivery of male to female gametes and the process of fertilization. A characteristic feature of the sexual process in gymno- and angiosperms is the independence of fertilization from water, in contrast to all plants and pteridophytes, in which fertilization occurs only in the presence of water, which is necessary for the movement of spermatozoids to the egg-cells. The main direction of the further evolution of the sexual process in the broad sense of this word consisted in the perfection of readjustments ensuring fertilization in the new conditions. The perfection of readjustments ensuring fertilization in the new conditions presented the main direction of the further evolution of the sexual process in gymno- and angiosperms.

Implementation of the functions of male and female gametophytes in the new conditions inevitably caused the structural changes in their organization, aimed at optimizing the passage of the fertilization process and determined the modern specificity of the embryo sac and pollen grains. On the basis of extensive comparative embryological studies of angiosperm species, classifications of types of ovules, micro- and macrosporangia, male and female gametophytes, fertilization, embryo- and endospermogenesis have been developed, and controversial issues of their emergence and evolution are actively discussed [4–16]. It must be emphasized that internal structures of the generative organs of angiosperms arose as a result of complex morphogenetic transformations of the homologous structures of the previous forms. Cognition of the laws of these transformations, i.e. becoming and further evolution of embryological structures, is the task of the evolutionary embryology of plants, in particular angiosperms. In this mini-review, we have made an attempt to discuss the evolutionary transformations of the embryological structures of angiosperms in the light of the main directions of the evolutionary process – aromorphosis, allomorphosis, specialization and reduction.

2. Aromorphosis

The currently accepted concept of aromorphosis [1, 2, 17, 18] includes the appearance in the body of such changes of a progressive order that are not strictly limited to any particular environment and, thus, raising the organization of the organism to a higher level, allow it to capture new, often very different areas for its habitat. Assignment of the emergence of angiosperm female and male gametophytes to phenomena of the aromorphosis type emphasizes the higher level of their development in comparison with the previous ones. The specific female gametophyte of angiosperms and double fertilization were new progressive steps, surprisingly precisely coordinated with each other, in the organization, first of all, of the plant female generative sphere and the implementation of one of the cardinal biological processes – fertilization. Polyploid endosperm, formed as a result of the fusion of sperm with the polar nuclei of the central cell of the embryo sac, is the final stage in the conjugated chain of aromorphoses of the internal structures of the generative organs of angiosperms. Deep rearrangements in the organization of the angiosperm female gametophyte, embryo sac, in comparison with the female gametophyte of archegonial plants, and primarily gymnosperms, is quite natural. We do not consider in detail the existing hypotheses about of the embryo sac origin and the directly related question about the origin of angiosperms, referring the readers to excellent reviews [4, 5, 9, 10, 12, 18–27].

We only note that Porsch's [28], and Favre-Duchartre's [29], archegonial hypothesis is based on the recognition of the embryo sac homology with the female gametophyte of gymnosperms. Coulter [6] expressed the idea of gradual reduction of an archegonium, which begins in gymnosperms and ends in angiosperms with the complete disappearance of its wall, from which only the reproductive structure remains, an egg cell. Fagerlind [30, 31] considered the elements of an embryo sac as the initial stages of the female gametophyte development in *Gnetum*, in fact, before the formation of archegonia. A clearly formulated hypothesis of the neotenic origin of an embryo sac was proposed by Romanov [32] and Takhtajan [9, 33]. According to these concepts, the maturation of gametes occurs before the formation of archegonia as a result of terminal abbreviation (shortening) of the gametophyte development. "Like any neoteny, a sharp shift in the reproductive phase (in this case, the differentiation of the ovule) to an early stage of development, at the latest after the third division of the macrospore nucleus, occurs together with the terminal abbreviation of development, that naturally excludes the archegonium formation" ([33], p. 116). Thus, a terminal abbreviation with the subsequent deviation of the last phases of development is clearly manifested in the embryo sac emergence. According to Kollman [34], who proposed this term back in 1884, the concept of "neoteny" (literally "retention of a youthful form)" means the preservation of juvenile characteristics in a sexually mature organism. Neoteny in organ development is also considered as a part of paedomorphosis (literally "child-shape" including neoteny (reduced rate) [8, 9, 35, 36], included in the broader concept of "heterochrony", which can be defined as change to the timing or rate of developmental events, relative to the same events in the ancestor [8, 11, 15, 37, 38] idea of the important role of neoteny in the emergence of the newest characters at the early stages of ontogenesis and, thus, more progressive taxa in evolution finds its development in the views of Takhtajan [10, 11] on macroevolution as a result of changes in developmental timing (heterochrony or heterobatmy).

The male gametophyte (pollen grain) of angiosperms has emerged on the same regularities as the female one – the onset of the phase of puberty at the early stages of gametophyte development, even before the formation of an antheridium, by the basal abbreviation of the gametophyte development of the previous forms [33] (**Figure 1**). As a result of the pistil formation, pollen germinated on the stigma being not protected from the direct influence of the environment [12, 39–44]. Pollen grains of gymnosperms germinate in a specific pollen chamber and further development of the male gametophyte occurs over a relatively long period of time [45–47]. Germination of angiosperm pollen grains on the stigma should have been carried out for a much shorter time [48]. So, only the final stage of the male gametophyte development could take place after pollination – the formation of gametes that occurs in plants with the two-celled type of pollen grains. In plants with the three-celled type of pollen grains, the entire development of the male gametophyte ends in the microsporangium. In general, the time interval between pollination and

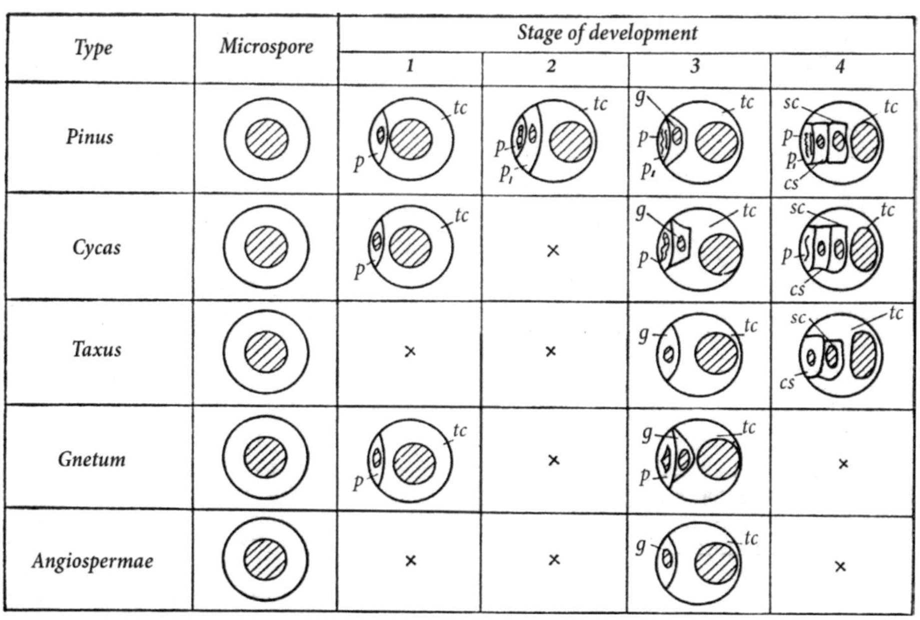

Figure 1.
Comparison of developmental stages of a male gametophyte in seed plants. X –stages missing from ontogenesis in a result of abbreviation; p, p₁ – prothallial cells; tc – tube cell (siphonogenic); g – generative cell; cs – cell-stalk; sc –spermatogenous cell [33].

fertilization in gymnosperms is much longer as compared to angiosperms. It ranges from 10 h to more than 12 months (fertilization occurs in the spring or summer of the next year), except for *Gnetum* and *Ephedra*, which have a fertilization interval of 6–8 days and 10–36 h, respectively [46–48]. Accelerated pollen tube growth rate in angiosperms is considered as a critical innovation that preceded the origin of the closed carpel, long styles, and multiseeded ovaries [40].

Angiospermic plants have a shorter reproductive cycle than gymnosperms [12, 39, 45]. An important step in the onset of a rapid reproductive cycle is to shorten of the progamous phase [12, 22, 42, 43]. The duration of the progamous phase is usually associated with the rate of the male and female gametophyte development. In general, well-hydrated and metabolically very active pollen germinates quickly (often within minutes) [49], but also rapidly loses viability [50]. Active pollen dehydration provides a longer functional life, but slower germination and leads to large fluctuations in the duration of the progamous phase [42, 43].

The syngamic phase in flowering plants can also vary considerably in duration. In a number of Asteraceae species, for example *Taraxacum koksaghyz* and *Crepis capillaries*, syngamy (fusion of sperm with an egg) and triple fusion (fusion of sperm with the polar nuclei of the central cell of the embryo sac) were observed within 15–30 min and 40–50 min after pollination, respectively [51]. There is a very short progamous phase of the fertilization process (from the moment the pollen gets the stigma to the entry of a pollen tube into micropyle of the ovule) in cereals, for example wheat – 15 min [52]. In *Amborella trichopoda*, syngamy and triple fusion could be observed within 24 h [41], in *Consolida arvensis*, 3–12 h after pollination [53] that is possibly due to the large number of ovules in the ovary and their non-simultaneous maturation. In most Liliaceae, especially in Lilium species, the duration of pollination–fertilization is very long, for example, the progamous phase continues about three days and the phase of syngamia lasts about four days [54]. One of the reasons for the delay in the rate of fertilization may be heterochrony as a shift in the synchronous development of male and female gametes [55].

3. Allomorphosis

According to Schmalhausen [2], allomorphosis is the most common type of evolution, in which the organism does not experience any significant complication of its organization, and the relationship between the organism and the external environment generally retains the same character. The term "allomorphosis", as noted by Schmalhausen [2], closes to idioadaptation of Severtsov [1] that provides the emergence of various adaptive forms within the same level of organization but it seems to us broader and therefore we used it to consider the evolution of the internal structures of generative organs of angiosperm plants. We believe that their subsequent changes at the level of allomorphosis emphasize the importance of new correlative connections as a result of the emergence of angiospermy.

That is why, allomorphosis is most clearly manifested after the next aromorphosis in the evolution of various plant groups. The data of comparative morphological and ontogenetic studies show an essential role of abbreviation (mainly basal and terminal, less often medial), deviation (terminal and medial), less often prolongation (terminal) in the rearrangements of the internal structures of the generative organs, relating to allomorphosis. In our opinion, the emergence of the existing diversity of a number of embryological structures, in particular various types of macrosporangium and microsporangium wall formation, *Allium*- and *Adoxa*-types of bisporic and tetrasporic embryo sacs, types of endosperm, etc., should be attributed to phenomena of the allomorphosis type.

3.1 Macrosporangium

The emergence of an ovule in seed plants must undoubtedly be attributed to aromorphosis. With the formation of integuments, a macrosporangium found itself in the new conditions, which created the preconditions for its further evolution in certain directions [10, 12]. Evolution of a macrosporangium at the level of phenomena such as allomorphosis took place in both gymno- and angiosperms. In the latter, the range of changes in the macrosporangium structure turned out to be much wider due to the occurrence of angiospermy, the largest aromorphosis, and the creation, as a result, of new correlative connections [56–61]. The classification of angiosperm macrosporangia, proposed by us [60] is based on the theory of the tunic and corpus of the structure and functioning of the vegetative shoot growth point (cited from [62]). According to this theory, a vegetative shoot growth point consists of one or more peripheral layers, cells of which divide mainly anticlinally. This initial growth zone is called the tunic. Tunic layers surround a group of cells dividing in different directions and representing an another initial growth zone – corpus. The tunic and corpus are considered as two interdependent growth zones not associated with certain anatomical and morphological areas of the shoot and leaves. Cells of the tunic outer layer divide only in the anticlinal direction and form an epidermis. The organization of the tunic – corpus is also present in the generative meristem – the meristem of an inflorescence, flower, and the generative meristem is mostly characterized by a two-layer tunic.

In the course of evolution, there was a transition from the structured apical meristem of ferns and gymnosperms with one or more initials to the stochastic meristems of the layered structure of angiosperms. This created conditions for the formation of a pool of meristematic cells, which could compete with each other for the creation of cell clones [63–65]. The layered organization (corpus-tunic) of the corpuscular meristem contributed to the relative isolation of these clones of cells and influenced the spread of mutations [66–68]. Mosaicity and instability of cell populations of the stem apex were apparently of key importance for the evolution and wide diversity and distribution of angiosperms. The emergence of a generative meristem could also be induced by mobile genetic elements [69]. These mechanisms allowed angiosperms to quickly get rid of harmful mutations, preserve genetic variants for a long time and in case of environmental threats, and cells carrying adaptive mutations to win the competition to provide a wide variety of structures of reproductive organs.

The classification of macrosporangia proposed by us [60] is found on the structural features of a nucellus, namely: a number of central cells of the tunic second layer of an ovule rudiment, a number of differentiating archesporial cells, the division intensity of vegetative, and in some cases, sporogenous cells of the tunic second layer, and time and place of the emergence of integuments. Two groups of macrosporangia are distinguished (**Figure 2**): (A) an undifferentiated ovule primordium consists of cells derived from the two tunic layers and a corpus layer. (B) An undifferentiated ovule primordium consists of cells derived from one tunic outer layer and the placenta subepidermal layer [70].

Based on the ideas about the direction of macrosporangium evolution in archegonial and angiosperm plants, we consider a large macrosporangium with a multilayered wall, multicellular archesporium and a well-expressed parietal tissue as a primary type I (*Casuarina*-type) [33, 51, 56, 60, 70]. Macrosporangia of II-VIII types are recorded as derivatives of type I which results the tripartite abbreviation, mainly basal, medial and terminal (according to Takhtajan [33]). The most striking example of basal abbreviation is the development of macrosporangium types VI (*Achillea*), VII (*Antennaria*) and VIII (*Himantoglossum*), where the initial stages of

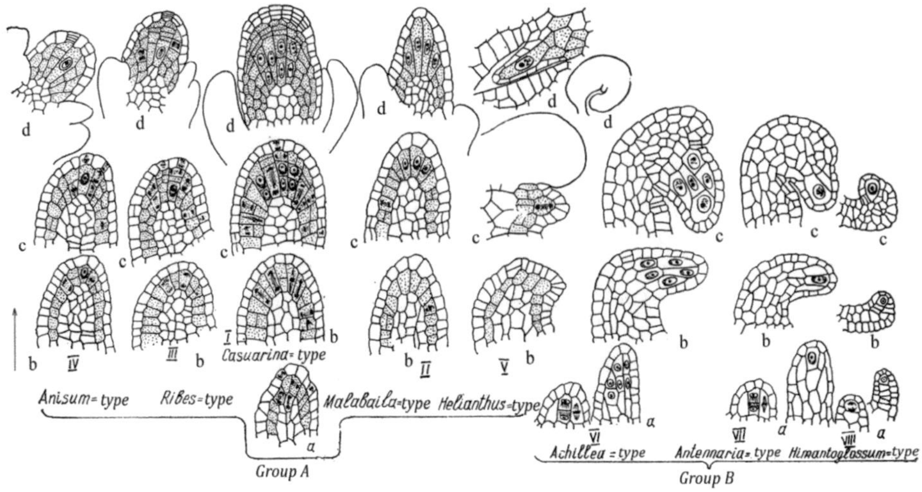

Figure 2.
Main types of macrosporangia in angiosperms: (a–d) successive stages of macrosporangium development [60].

the ovule primordium formation almost completely disappear. In macrosporangia of these types, archesporial cells differentiate from derivatives of the placenta subepidermal layer after two or three periclinal cell divisions. Medial abbreviation is widespread under development of types II – IV. For example, primary archesporial cells, differentiating from the central cells of the tunic second layer of the macrosporangium primordium, as well as protoderm cells of its apex do not divide under development of a type II (*Malabaila*). Division of the primary archesporial cell also falls out during the formation of type IV (*Anisum*). The number of divisions of somatic cells of the tunic second layer was also reduced during the development of type II and IV. One or another derivative type of a macrosporangium does not depend on the type of ovule (atropic, anatropic, hemitropic, campylotropic, amphitropic), as well as the position of the ovules in the ovary – epitropic, apotropic and pleurotropic. Thus, evolution of angiosperm macrosporangium proceeded in the direction of reduction of its vegetative and sporogenous parts, and both directions might not coincide in time. The most advanced types of macrosporangia should be considered IV (*Scopolia*- and *Vincetoxicum*- subtypes), VII (*Antennaria*-type) and VIII (*Himantoglossum*-type) characterized by a single-layer wall, unicellular archesporium and the absence of parietal cells [60]. It should be noted that in nature there are undoubtedly intermediate forms between the main types of macrosporangia.

3.2 Microsporangium wall

The medial and terminal abbreviations, in our opinion, are clearly traced in the formation of an microsporangium wall during angiosperm evolution and expressed in the reduction of a number of cell divisions derived from the parietal layer. In a number of angiosperm species, the microsporangium wall consists of five or more layers; as a result of cell division in the parietal layer, two central layers arise. Periclinal cell divisions of the two central layers lead to the formation of the tapetum, two middle layers, and an endothecium. A further increase in the number of layers of the microsporangium wall occurs mainly due to periclinal cell divisions of one or two middle layers. Additional middle layers can also result from cell division of the endothecium or tapetal layer. Davis [61] considered the last type of wall formation as the initial type and gave it a corresponding name – the basic type. The centrifugal type of microsporangium wall formation, since the time of Warming [71], is known for the fact that as a result of cell division of the parietal layer, the tapetum and the central layer are formed, from which the middle layer and endothecium arise. In the centripetal type, as a result of cell division of the parietal layer, endothecium and the central layer are formed, from which the middle layer and tapetum arise. Thus, the wall in those cases consists of four layers – the epidermis, endothecium, middle layer and tapetum. The centrifugal type of formation of the microsporangium wall is known mainly for dicotyledonous species, which led to Davis's [61] proposal to designate it as a dicotyledonous type. The centripetal type, monocotyledonous according to the terminology of Davis [61], is more common in monocotyledonous species. In the reduced type, the tapetum and endothecium are directly formed from periclinal divisions of the cells of the parietal layer, the middle layer is absent (**Figure 3**). There is no objection to the position of Davis that dicotyledonous (centrifugal), monocotyledonous (centripetal) and reduced types of microsporangum wall formation are derivatives of the basic type, in which a more massive microsporangium wall is formed as a result of a greater number of cell divisions. The emergence of various types of angiosperm microsporangia, as well as macrosporangia, can be attributed to phenomena such as allomorphosis.

3.3 Embryo sac

Such modus of evolution of the female gametophyte in angiosperms (it also occurs in *Gnetum*) [46], as the combination of stages of macrosporogenesis and the embryo

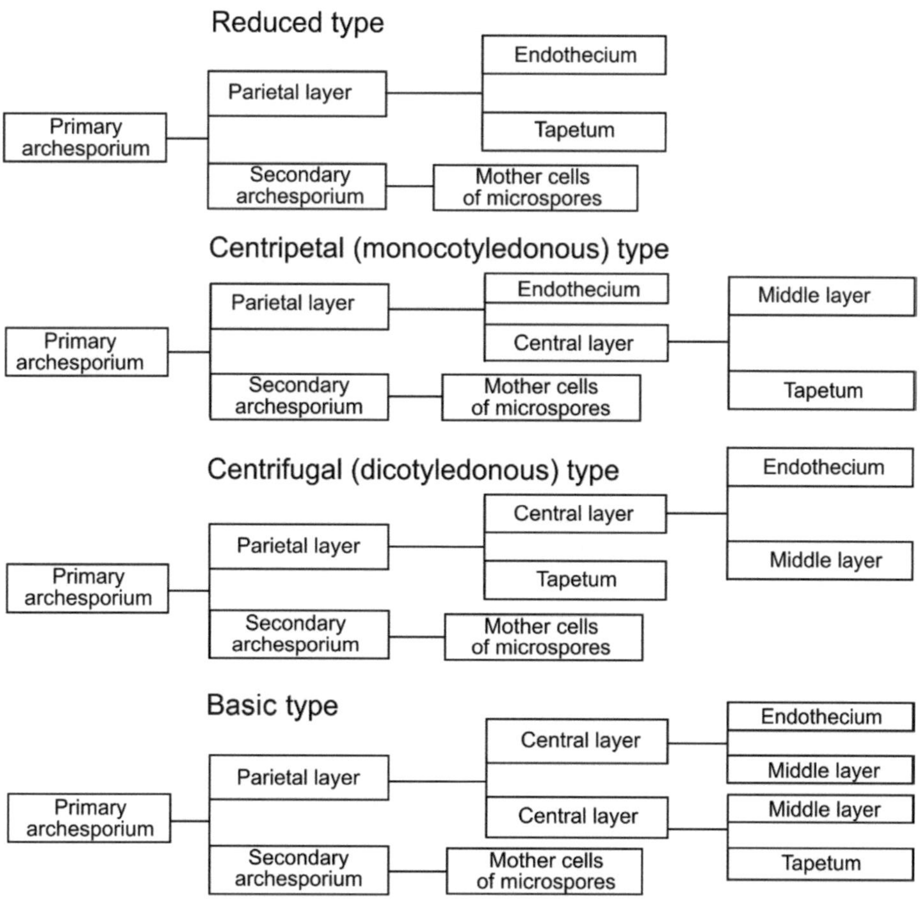

Figure 3.
Types of microsporangium wall formation [61].

sac development, clearly reveals in the origin of bisporic and tetrasporic embryo sacs [32, 72]. So, the tetranuclear embryo sac is accomplished after the fourth (*Allium*-type) and third divisions (*Adoxa*-type). Nevertheless, it is more correct, in our opinion, to speak not about the loss of one or another stage, but about overlapping the stages. The *Oenothera*-type monosporic unipolar embryo sac could originate from the normal type, in particular through terminal abbreviation (only two divisions of a macrospore nucleus) and medial deviation (after the first division of a macrospore nucleus, two nuclei are located at the micropylar end of the embryo sac) [32, 72]. The emergence of a peculiar tetrasporic embryo sac of the *Fritillaria*-type is possible to be a medial deviation. Medial and terminal deviations are also traced in the development of tetrasporic embryo sacs of the types *Plumbago, Plumbagella, Tulipa tetraphylla*, and *T. eriostemon*. Tetrasporic embryo sacs of the *Drusa, Peperomia*, and *Penaea* types arise through a overlapping the medial and terminal deviations and terminal prolongation [12, 32, 72]. The structural diversity of bi- and tetrasporic types may be the result of instability and competition between macrospores participating in their ontogenesis [73]. This competition within embryo sacs can be expressed as eggcell production by descendants of more than one megaspore, or suppression of the descendants of all but one megaspore. Both phenomena evolved several times, forming obviously an amazing variety of the structure of the embryo sacs [73].

3.4 Endosperm

The triploid endosperm produced by most monosporous gametophytes is likely descended from the ancestral diploid endosperm and is considered a key adaptation of the most successful flowering plant lines [10–12, 73–75]. The greater importance attached to the duplicated maternal genome in the triploid endosperm could have contributed to a more efficient allocation of resources in the embryo sac, since it reduced the influence of competition for maternal resources unrelated among paternal genomes of endosperm [73]. At the same time the actual ploidy of the endosperm nuclei can vary widely depending on the type of embryo sac, the involvement or depression of the lower polar nucleus, the stage of endosperm development and other events [32, 74–80].

The differences between three types of endosperm – nuclear, cellular and helobial – are mainly observed in order variations of mitosis and cytokinesis at the initial stages of endosperm development [51, 58, 61, 74–80]. Subsequent stages of endosperm development, such as synthesis and accumulation of reserve nutrients, its presence in mature seeds or its resorption by an embryo do not show any direct connection with its types. The developmental features of cellular and helobial endosperm have no analogues in archegonial plants. Regardless of the type of endosperm development, its role in the formation of the embryo, seed, and fruit is basically the same, which has been shown by numerous studies of the embryology and biology of seed and fruit maturation in a number of cultivated angiosperm species under natural conditions and experiments [74–80]. Attempts to compare this or that endosperm type with the shape and size of embryo sacs, as well as the rate of embryogenesis in cultivated and wild angiosperm plants have not yielded positive results [80]. This gives us the basis to consider the types of endosperm at the level of allomorphosis phenomena as various kinds of basal deviations, hereditarily fixed in the further evolution of angiosperm taxa.

4. Specialization

In plants, the division of labor between functionally specialized modules occurs at all levels of biological organization. Division of labor is favored by three factors: positional effects that predispose modules for one of the tasks, accelerating performance functions, and synergistic interactions between modules [81]. Under specialization of internal structures of generative organs of angiosperm plants, we consider the changes which do not lead to a visible increase in the level of overall organization. Specialization directly works to optimize the implementation of the main functions of a given structure or its part [81], it manifests itself at the cellular and tissue levels and usually not directly related to the ecological or biological characteristics of species and, therefore, it is found in different taxa. As examples, we consider various types of the tapetal layer of a microsporangium, differentiation of the integumental (integumentary) tapetum, and various kinds of haustoria.

4.1 Microsporangial tapetum

The morphological features of the tapetal layer indicate its active physiological state associated with its important functions – to supply the sporogenic tissue with nutrients and produce the specific substances necessary for the normal course of microsporogenesis and the male gametophyte development [12, 51, 81]. Despite the type of microsporangium wall formation, the tapetal layer is always its innermost layer. There are two main types of a tapetal layer – secretory or parietal and amoeboid or periplasmodial [82, 83]. Secretory tapetum cells lose their individuality not earlier of the formation of microspore tetrads or at the subsequent stages of male gametophyte development, up to two-celled pollen grains. Periplasmodial tapetum cells lose their individuality during the prophase of meiosis of the microspore mother cells. Protoplasts are joined, forming a typical, or true, periplasmodium with numerous nuclei. Once the tetrads of microspores have split up, periplasmodium fills the entire cavity of the microsporangium, surrounding microspores, and later pollen grains. The amoeboid type of tapetum is sometimes considered as a of the secretory type, as its formation begins at the stage of microspore tetrads or mononuclear microspores, and a part of the cells can retain their individuality [82]. Recent studies indicate some archaic features in the behavior of cells of the periplasmodial tapetum at the beginning of microsporogenesis. For example, some lilies in anthers with a large number of microsporocytes develop a competitive cytomictic relationship between microsporocytes and tapetum cells [84], probably for the space in the anther [85]. These processes are apparently regulated by microsporocytes [86], providing the necessary tissue homeostasis in the developing anther.

It is believed that the periplasmodial tapetum repeatedly appeared in early angiosperms, especially in monocots [87]. The evolutionary tendencies of the tapetum may be directed towards even closer contact with spores / pollen grains. In this respect, three evolutionary trends are distinguished: (1) invasion of tapetal cells between spores, (2) loss of tapetal cell walls, and (3) increased feeding through direct contact in narrow anthers [83]. A common feature of both tapetum types is polyploidization, undoubtedly associated with its secretory function, in particular the secretion of sporopollenin precursors. The secretory function of the tapetum is also confirmed by its ultrastructural organization, which is similar to that of actively secreting cells. The differences between the secretory and periplasmodial types are mainly in the time and degree of cell wall lysis resulting in the formation of a mostly false or typical periplasmodium. In general, the affinity in the cytoplasm ultrasructure, behavior of nuclei, and the physiological activity of the periplasmodial and secretory types gives us reason to consider them only as a display of the structural specialization of the innermost layer of the microsporangium wall.

4.2 Haustoria

Extreme variety is characteristic for haustoria – haustoria of synergids, antipodal cells endosperm, and suspensor [88, 89]. The number of genera of dicots and monocots reported to develop haustoria was 305 and 51, respectively [89]. Twelve dicot families had six, or more, genera with reported haustoria differentiating within ovules or seeds: Rubiaceae (6 haustoria), Proteaceae (6), Campanulaceae (7), Verbenaceae (9), Santalaceae (10), Asteraceae (11), Solanaceae (13), Cucurbitaceae (16), Loranthaceae (17), Fabaceae (23), Acanthaceae (25), and Scrophulariaceae (40) [89]. These twelve families constituted 60% of the total number of dicot genera reported as having haustoria [89]. Only three of the 16 monocot families had six or more genera with haustoria: Orchidaceae (15); Poaceae (7) and Araceae (6), which accounted for 51% of the total documented monocot genera differentiating haustoria [89].

Synergid haustoria are formed as a result of the elongation and increase in a volume of the upper part of synergids, which penetrate more or less deeply into micropyle; the longest synergistal haustorium has been described in *Quinchamalium chilense* (**Figure 4a–e**) [90], *Olax stricta* and *O. wightiana* [91, 92] and other parasitic angiosperms [92]. The female gametophyte of these species elongates and grows beyond the ovule into the base of the stylar canal. A lateral caecum develops close to the chalazal end, and extends in the basal region of funiculus [88]. Cotyledonous haustoria are described in certain monocotyledonous families such as the Liliaceae [93], the Arecaceae [93], the Palmae [94] and the Poaceae [95], endospermal haustoria are described in certain species of Cucurbitaceae, Fabaceae, Opiliaceae, Euphorbiaceae [89]. The apex of the antipodal haustorium in these species passes through the funiculus, reaching the placenta apex, where it forms branches that penetrate the tissue of the placenta. The nuclei are hypertrophied, contain numerous nucleoli, and sometimes divide, resulting in five to seven nuclei. In another species the chalazal part of an embryo sac becomes almost three times longer than the micropylar part due to development of the antipodal haustorium [96]. The antipodal nuclei do not organize into independent cells so that there is an antipodal chamber with three nuclei. The tip of the antipodal chamber elongates, passes through the funiculus, reaches the placenta, and branches therein. The antipodal nuclei become hypertrophied and, sometimes, 5–7 nuclei are delimited by a partition wall [90].

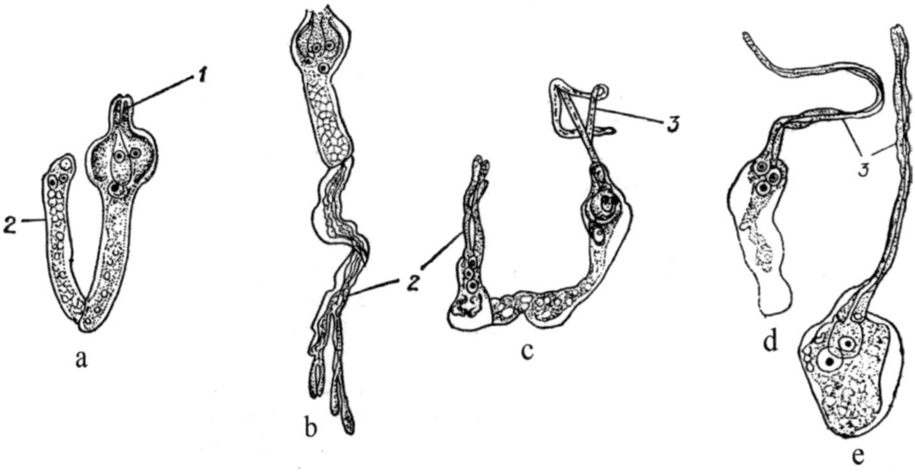

Figure 4.
Mature embryo sacs of Quinchamalium chilense with antipodal haustoria (a and b), antipodal and synergidal haustoria (c); (d and e) fertilized embryo sacs with synergidal haustoria. 1 – synergid, 2 – antipodal haustorium, 3 – synergidal haustorium [90].

The most widespread and peculiar structures are of endosperm and suspensor haustoria, that is quite understandable, since they strengthen the influx of nutrients to a developing embryo. Endosperm haustoria are the most diverse in their structure and developmental features in species with cellular and helobial endosperm types. In the nuclear type of endosperm, the haustoria mainly arises from the chalazal end of the embryo sac. In most cases, they are short and remain at the coenocyte stage. In angiosperms with endosperm of cellular and helobial types, haustoria can arise at the chalazal and micropylar ends of the embryo sac, as well as at both. In *Opilia amentacea* (Opiliaceae) [97], the division of the primary endosperm nucleus leads to the formation of a micropylar and a long tubular chalazal chamber. The development of the endosperm in *Aeginetia indica* (Orobanchaceae) is of the cellular type [98]. In addition to such haustoria, several secondary lateral haustoria can be formed. In a number of species of the family Scrophulariaceae [58], well-developed and aggressive haustoria of all three types can arise simultaneously, thus forming a very effective absorbing system [58].

Suspensor haustoria can be one-, two-, and multicellular, as well as more or less branching, contain one or two nuclei, or represent a well-developed coenocyte with numerous nuclei. Suspensor haustoria are quite common in embryos of several taxa: Cuscutaceae, Leguminosae, Podostemaceae and Tropaeolaceae [88]. For example, the branches of a suspensor haustorium in *Tropaeolum majus* (Tropaeolaceae) pass the integuments and funiculus reaching the site, where a vascular bundle enters into seed juncture. Polyploidy in suspensor cells is quite common. For example, in *Phaseolus vulgaris* (Fabaceae) [99], the basal cell of the 2-celled proembryo produces a multicellular suspensor consisting of mononuclear giant cells with polytene chromosomes. Due to endoreduplication the nuclei exhibit high polyploidy (up to 4000–8000n) [100].

4.3 Integumental tapetum or endothelium

Differentiation of the internal epidermis of the integument into integumental tapetum (this name was given to this layer due to its similarity the microsporangium tapetum) or endothelium also can attribute to specialization display. It should be noted that integumental tapetum differentiate not in all species, a mature embryo sac of which is surrounded by the integument after nucellus destruction. Presence of endothelium is normally restricted to tenuinucellate and generally unitegmic ovules [101, 102]. Endothelium occurs preferably in the dicotyledons, rarely in some monocots [14, 56, 101, 103]. The integumental tapetum is most clearly expressed in many families of Santalales, Lamiales, Solanales, Oxalidales, Malpighiales (in particular Linaceae, Lobeliaceae, Asteraceae Convolvulaceae, Rubiaceae, Oleaceae families), as well as in some families of Pandanales, Poales, Asparagales (Orchidaceae) [101–106]. So, typical endothelium was described in species of *Hieracium*, *Taraxacum*, and *Helianthus* [101, 107, 108], *Bellis perennis* [109], *Chondrilla* ovules, both in sexual and apomictic plants [101, 108], species of *Abeliophyllum* [110], *Petunia* [111], *Arabidopsis* [112] and in others.

The functional significance of the endothelium in angiosperms is not completely clear, but it likely plays a role in regulating the development of the embryo [104–106]. The integumental tapetum can play the role in the transfer of nutrients from the integument to the embryo sac, as well as in the production of a number of enzymes (diastase, protease, etc.), which convert nutrients into a form assimilable for the embryo sac [103, 109, 113, 114]. Like the perisperm, this well-developed endothelium may be a morphological manifestation of maternal control over development following the evolution of double fertilization [106, 115]. The presence of an endothelium has traditionally been regarded as a relatively advanced feature of seed development because it is widespread among asterid angiosperms [101, 105, 106]. Among the lineages that diverged early in angiosperm evolution an endothelium has been recorded only in extant *Lactoris* (Lactoridaceae, Piperales) [116] and *Nymphaea thermarum* (Nymphaeales) [115]. Thus, the integumental tapetum performs a glandular-secretory function, although it is also considered as a barrier preventing the transport of nutrients, which come from the chalazal part of an ovule to an embryo sac into integument. Thus, the integumentary tapetum is a specialized tissue that forms at the border of gametophyte and sporophyte in usually reduced ovules. It can serve a nutrition function, although it is also seen as a protective barrier or as tissue exercising control over the developmental embryo by the sporophyte [14, 106, 108, 109, 115, 116].

4.4 Antipodals

The antipodal apparatus is a group of a wide range of specialized cells. In mature and fertilized embryo sacs, the shape, size, number of antipodal cells and their nuclei essentially vary. At the same time, antipodals represent temporary formations in many angiosperm species and are actually absent in the embryo sac during fertilization. Here, we briefly review the structure and behavior of antipodal cells with haploid nuclei, which are formed in an embryo sacs of the *Polygonum*-type, which is the most widespread among angiosperms, as well as in embryo sacs *Allium*-, and *Adoxa*-types [32, 33, 35, 51, 56, 72]. The differences in a number and ploidy of antipodes in other types of embryo sacs are due to the peculiarities of their development, for example triploid antipodal cells in the embryo sacs of *Fritillaria*- and *Plumbagella*-types [32, 56, 72].

The division of antipodal cells usually begins during the maturation of the embryo sac and a number of antipodes can increase significantly, in particular in many species of Poaceae family, e.g. up to 300 in *Sasa paniculata*, a bamboo [117]. In *Aconitum vulparia* the giant endopolyploid antipode nuclei are lobed and seem to be at their most active state [118]. An antipodal complex consisting of a different number of cells (always more than three, and often up to 150) was described in species of Dipsacaceae, Gentianaceae, Pandanaceae, Rubiaceae, Typhaceae and Sparganiaceae families [14, 51, 56, 58, 119].

Antipodal cells can remain mononuclear or become bi- and multinuclear as a result of mitotic division of their nuclei without subsequent cytokinesis. For a number of angiosperm species, polyploidization of antipodals is known and they may be present in the embryo sac for a considerable time after fertilization, up to the stage of the globular embryo. As an example, antipodales of many species in the Ranunculaceae family grow thick, become hypertrophied glandular formations at the stage of nuclear endosperm and persist for a long time in the embryo sac, up to the stage of the multicellular embryo and cellular endosperm, in which reserve nutrients accumulate. Polytene chromosomes are described in antipodal cells, in particular in the species of Papaveraceae, Poaceae and Asparagaceae families [120–123]. In maize the antipodal cells continue to divide during embryo sac maturation reaching a final number of 20–100 cells with one to four nuclei each [124]. Antipodal cells in maize also have high metabolic activity and nutritive function (high sucrose synthase activity), involved in auxin biosynthesis and signaling into embryo sac [125, 126]. Interestingly, the antipodal cells in species of lily (*Fritillaria*-type embryo sac) are characterized by a definite degree of affinity to egg apparatuses, e.g. a special type of specularity (symmetry) attracting an additional pollen tubes in chalazal compartment of embryo sac [54]. Multiple antipodal cells are common in Ranunculales, Papaverales, Rhamnales and Malpighiales [56, 58, 118, 119]. In sympetalous groups large antipodal cells are recorded from some Comales and a few Lamiales but are otherwise rare [14, 56, 58].

The functional significance of antipodes, which persist for a long time in embryo sacs, is interpreted ambiguously. Antipodals are regarded as a physiologically active apparatus that play an important role in the nutrition of the embryo sac [127–129], being the center for the absorption, processing, and transport of nutrients from the ovule chalaza into embryo sac. The opposite point of view about the absence of vital functions in the development of the embryo sac in the antipodal apparatus is based on the strong degree of its variability and the concept of the rudimentary nature of antipodal cells [119]. Despite directly opposing points of view regarding the functions of antipodales in the embryo sac, there is generally a consensus about the antipodal complex as a specialized apparatus of the embryo sac, regardless of its passive or active participation in the metabolism of the embryo sac.

Thus, specialization of various embryological structures does not show any visible connection with the general level of development of one or another taxon, appreciating by a whole complex of characters. Parallelism, recurrence of occurrence and mechanisms of formation of embryological structures can be traced both in early and evolutionarily advanced taxons of angiosperms. The same types of a tapetal layer, similar features of the formation, structure and behavior of haustoria, antipodes and their nuclei are described in the families of angiosperms, occupying very different positions in the system, as well as in families rich in species with a wide ecological plasticity and in specialized families adapted to strictly defined habitation. At the same time, certain trends in development can be traced, indicating specialization associated with taxon-specific traits.

4.5 Specialization relevant to to plant ecology

In addition, we consider some examples of specialization more clearly related to the special features of plant biology and ecology in Podostemaceae, Loranthaceae, Olacaceae, Santalaceae and Viscaceae families. Podostemaceae family is a pronounced line of hydrophilic evolution [130, 131] and widespread in tropical countries. This family includes herbaceous plants, often of negligible size, growing in fast-flowing water and characterized by a high specialization of vegetative organs and their extreme diversity. The embryo sac in this family is distinguished as a special type - *Podostemon*, which is a variant of the *Allium*-type of embryo sac with only four cells as a result of terminal abbreviation [132–134] as abovenoted, formation of a false embryo sac and suspensor haustoria are considered as adjustment to provide nutritions of an embryo in the absence of endosperm. Organization of three-celled embryo sac in the species is the result of precocious degeneration of the central cell prior to syngamy. Lack of central cell before fertilization leads to absence of double fertilization and endosperm in the species, and this features reiterates the general cause of usual absence of double fertilization among the podostemads [135–137]. The nucellar plasmodium forms before fertilization in the Podostemoideae and Weddellinoideae but after fertilization in the Tristichoideae of the Podostemaceae. The timing in the formation of the nucellar plasmodium in either subfamilies varies and is species-specific. Also, the development of the female gametophyte is species-specific with limited life span of the central cell. The second male cell degenerates such that there is no endosperm rather nucellar plasmodium is thought to offer the nourishment to the developing zygote [138].

The Loranthaceae, Olacaceae, Santalaceae and Viscaceae families are represented by woody and grassy forms, often leading a parasitic or semi-parasitic life style on tree branches or roots [130, 139–141]. Species of these families are mostly distributed in tropical and subtropical areas, only a few are found in the temperate zone. Overgrowth of an embryo sac beyond the nucellus and formation of lateral or chalazal haustorium penetrating into the funiculus and placental column was

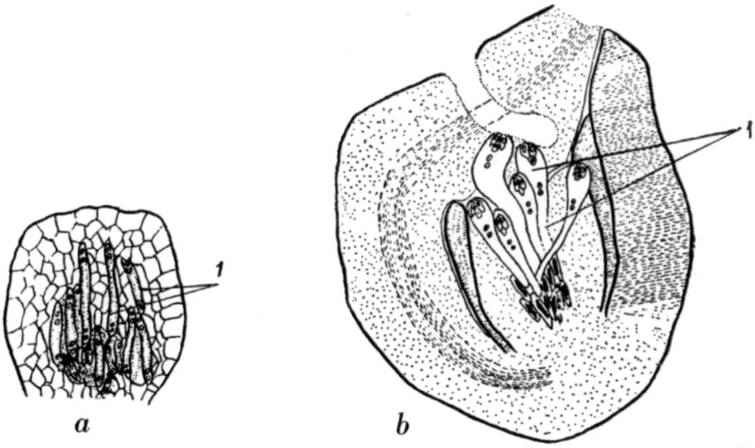

Figure 5.
Apical part of nucellus with embryo sacs at different stages of development (a) and ovule with mature embryo sacs in Cassytha filiformis (b). 1 – embryo sac [144].

reported for *Osyris wightiana* [89, 142]. Enlargement of an embryo sac by such a way that its apex can reach the style and stigma has been reported only for the Loranthaceae family, e.g. the embryo sacs reach the stigma in *Tapinostemma acaciae* [143]. The female gametophytes of this species grow from their tips and compete over hundreds of micrometers to reach the apex of the nucellus and the site of fertilization [143]. In *Helixanthera ligustrina*, the apex of an embryo sac is surrounded only by the epidermis of the stigma. Several embryo sacs, a number of which can reach more than six, develop in the ovule [88, 144], from one to four embryo sacs penetrate the funiculus in *Cassytha filiformis* (**Figure 5a** and **b**) and *C. pubescens*. Intense growth of embryo sacs, in some species of Loranthaceae family, in which the tops of the embryo sacs with the egg apparatus reach the stigma, is assumed to facilitate the fertilization, especially in cases of the micropyle absence.

5. Reduction

The term “reduction” is widely used in discussions about the evolution of plant embryological structures – reduction of gametophytes, integuments, nucellus, and embryos etc.). Nevertheless, the meaning enclosed in it is ambiguous in different cases. Least of all, this term is used in the literal sense – aplasia or loss of one or another trait. This is quite understandable, since gamete formation is the final stage in the development of gametophytes. In this regard, the term “reduction” as applied to gametophytes most likely has a synonym “abbreviation of development” or neoteny [9, 14, 32]. In Romanov’s opinion, the simplification of the structure of angiosperm gametophytes is only a morphologically simple expression of a complex chain of biochemical and physiological reactions. A female gametophyte is also preserved in apomictic plants, often transforming from haploid to diploid [32].

Various forms of reduction of the ovule structure and embryogenesis, often constituting a fairly clear reduction series, are noted among the epiphytic, saprophytic and parasitic species of a number of angiosperm families – Scrophulariaceae, Santalaceae, Balanophoraceae, Cynomoriaceae, Lennoaceae, Hydnoraceae, Loranthaceae, Myzodendraceae, Rafflesiaceae, Lauraceae, Cuscutaceae, Pyrola ceae, Burmanniaceae, Gentianaceae, Lobeliaceae, Polygalaceae, Triuridaceae, Corsiaceae, Petrosaviaceae, Orchidaceae, and Viscaceae [9, 140, 145–148]. An example, two archesporial cells differentiate directly in the subepidermal layer of the young column located in the center of the ovary and representing the placental-ovule complex on the opposite side in *Arceuthobium minutissimum* (Viscaceae), which is the smallest woody parasitic plant losing morphologically pronounced ovules [140, 145, 146]. In *Aeginetia indica* the ovules are extremely small, tenuinucellate, anatropous and unitegmic. The hypodermal archesporial cell functions directly as the mother cell of the megaspore, without cutting off the parietal cell. The global embryo is not differentiated, seeds are produced in abundance, but many of them are empty and devoid of an embryo or endosperm [98]. Among Santalales holoparasites, *Helosis* (Balanophoraceae) is intermediate in the reduction series of its floral organs [147]. The female gametophyte, named *Helosis*-type, is a bisporic four-celled embryo sac, provided with a typical egg apparatus and a uni-nucleated central cell. Fertilization was not observed, yet a few-celled embryo and cellular endosperm developed. The archesporial cells are oriented obliquely to the longitudinal axis of the column and are separated from each other by two or four rows of cells, they become directly the macrospore mother cells. The placental-ovule complex reaches its extreme reduction in species of the genus Balanophora, which are widespread in tropical, less often subtropical regions, and characterized by a very strong simplification of all organs. Archesporial cells differentiate directly into ovary tissue [9, 140, 141, 145]. In general, evolution of numerous parasitic and symbiotic organisms is known to entail simplification rather than complexification [149].

Considering the patterns of evolutionary transformations of the reproductive structures of semi-parasitic and parasitic angiosperms, Terekhin [148] emphasizes the fact that the transition to a new way of nutrition and life during the vegetation period is carried out under conservation of the old way of sexual reproduction. This leads to an obvious complication of ontogenesis: the transition to development with metamorphosis and the development of the protosomal system. Two main directions of structural reduction were distinguished – general (reduction of the structure size, the ratio between the individual parts of the reducible structure does not change) and differential (differentiated loss of parts of the reducible structure, the ratio of parts changes). In the course of evolution, both directions of reduction are conjugated. The reduction of the embryogenesis in heterotrophic plants is carried out by means of terminal abbreviation, that is quite natural, since the reduction of embryogenesis, being a sharply expressed, special case of specialization, does not lead to general progressive changes in the organization of one or another structure.

6. Concluding remarks

In the last three decades, a new approach to understanding the evolution and development of organismic structure diversity has grown into a striking discipline called "evolutionary developmental biology" (EvoDevo) widely using advancements and ideas of molecular biology and genetics, including genomics, transcriptomics, and proteomics, as well as bioinformatics [18, 150–154]. At the same time, comparative plant morphology is quite rightly emphasized should be a valuable sub-discipline of EvoDevo [18, 153]. In the broad sense, EvoDevo comprises form and structure at all organizational levels, i.e. "the form and structure of whole plants, organs, tissues, cells, cell organelles, molecules, etc. and it continues to play a major role in evolutionary biology" [155]. The most striking example of using the traits of of the micromorphology of the androecium and gynoecium, seeds and fruits, as well as the types of microsporangia, pollen grains, ovules, male and female gametophytes, endosperm, etc. is a new version of the Takhtajan's phylogenetic system of magnoliophyta [11]. In his opinion, evolutionary taxonomy should be based not only on cladistics analysis, but also on the rates and modes of evolution of individual groups, the degree of their divergence and the level of specialization.

Evolutionary transformations of embryological structures in the general directions and modes of progressive evolution and specialization reveal also the nature of new correlative connections as a result of the indirect influence of the external environment and their changes in the process of evolution. In the evolution of embryological structures there are, on the one hand, complicated correlative connections, on the other hand, more clearly general cellular patterns, i.e. mechanisms of mitosis, cytokinesis, polarity. Such a statement associates primarily with the function of these structures, providing direct plant seed reproduction, and their position as internal structures of generative organs and, consequently, their greatest conservatism in comparison with plant vegetative organs.

It seems expedient to study the mutational and phenotypic variation of the internal structures of angiosperms and the specificity of natural selection of mutations in these structures, which ensure the species reproduction, in particular the stabilizing role of selection. Many mutational changes in embryological structures cannot become the basis for the emergence of stable types if they negatively affect fertility. The limitations imposed by the selection of only those mutations that do not reduce fertility are one of the reasons for the widespread parallelism in the evolution of embryological structures. The specialization of embryological structures at the tissue and cellular levels in the direction of optimizing the implementation of their functions leads to convergent similarity, which manifests itself at these levels in the form of tissue specificity.

It is quite clear that the use of the achievements of molecular biology and genetics should be very fruitful in solving the problems facing evolutionary embryology – the knowledge of the laws of the emergence and historic development of the internal structures of generative organs. Cognition of the gene families and gene regulatory networks underlying development of male and female gametophytes and double fertilization in extant plants will be new insights into understanding pathways of their reproductive evolution. Although the final solution to these issues is still far away, the establishment of new concrete truths at various structural levels of the organization of plant organisms, expanding the boundaries of knowledge and confirming the achieved or refuting it, inevitably raises new debatable questions in such a complex, but extremely interesting and inexhaustible problem as the problem of the history and evolution of plant forms.

Author details

Elizabeth L. Kordyum[1*] and Helen A. Kravets[2]

1 Institute of Botany, National Academy of Sciences of Ukraine, Kyiv, Ukraine

2 Institute of Food Biotechnology and Genomics, National Academy of Sciences of Ukraine, Kyiv, Ukraine

*Address all correspondence to: elizabeth.kordym@gmail.com

References

[1] Severtsov AN. The Main Directions of Evolutionary Process. Moscow–Leningrad: BiomedGIZ; 1934. 150 p. (In Russian).

[2] Schmalhausen II. Ways and Patterns of the Evolutionary Process. Moscow–Leningrad: USSR Acad. Sci. Publ. House; 1940. 231 p. (In Russian)

[3] Golenkin MI. Winners in the Struggle for Existence. Moscow: Moscow State University; 1927. 101 p. (in Russian).

[4] Arber EAN, Parkin J. On the origin of angiosperms. Bot. J. Linn. Soc. 1907; 38:29-80.

[5] Coulter JM. Relation of megaspores to embryo sacs of angiosperms. Bot. Gaz. 1908; 45:361-366.

[6] Coulter JM. The Evolution of Sex in Plants. Chicago: Univ. Chicago Press; 1914. 140 p.

[7] Eames AJ. Morphology of the Angiosperms. New York: McGraw-Hill; 1961. 518 p.

[8] Takhtajan AL. Patterns of ontogenetic alterations in the evolution of higher plants. Phytomorphology. 1972; 22:164-171.

[9] Takhtajan AL. Neoteny and the origin of flowering plants. In: Beck CB, editor. Origin and Early Evolution of Angiosperms. New York: Columbia Univ. Press; 1976. p. 207-219.

[10] Takhtajan AL. Evolutionary Trends in Flowering Plants. New York: Columbia Univ. Press; 1991. 241 p.

[11] Takhtajan A. Flowering Plants. Netherlands: Springer; 2009. 871 p. DOI: 10.1007/978-1-4020-9609-

[12] Stebbins GL. Seeds, seedlings, and the origin of angiosperms. In: Beck CB, editor. Origin and Early Evolution of Angiosperms. New York: Columbia Univ. Press; 1976. p. 300-311.

[13] Choi JS, Friedman WE. Development of the pollen tube in *Zamia furfuracea* and its evolutionary implications. Am. J. Bot. 1991; 78: 544-560. https://doi.org/10.1002/j.1537-2197.1991.tb15221.x

[14] Dahlgren G. Steps toward a natural system of the dicotyledons: embryological characters. Aliso: J. System. Flor. Bot. 1991; 13:107-165. DOI:10.5642/aliso.19911301.05

[15] Friedman WE, Carmichael JS. Heterochrony and developmental innovation: evolution of female gametophyte ontogeny in *Gnetum*, a highly apomorphic seed plant. Evolution. 1998; 52:1016-1030. https://doi.org/10.1111/j.1558-5646.1998.tb01830.x

[16] Endress PK. Links between embryology and evolutionary floral morphology Curr. Sci. 2005; 89: 749-754. https://www.jstor.org/stable/24111019

[17] Severtsov AS. Establishment of aromorphic organization: Causes and conditions. Paleontol. J. 2010; 44: 1518-1524. https://doi.org/10.1134/S0031030110120075

[18] Olsson L, Levit GS, Hoßfeld U. Evolutionary developmental biology: its concepts and history with a focus on Russian and German contributions. Naturwissenschaften. 2010; 97:951-969. DOI 10.1007/s00114-010-0720-9

[19] Favre-Duchartre M. Homologies and phylogenies. In: Johri BM, editor. The Embryology of Angiosperms. Berlin: Springer-Verlag; 1984. p. 697-734. DOI:10.1007/978-3-642-69302-1

[20] Krassilov V. The origin of angiosperms. Bot. Rev. 1977; 43:143-176 DOI:10.1007/BF02860852.

[21] Krassilov V. The origin of angiosperms: New and old problems. Trends Ecol. Evol. 1991. DOI: 0169-5347(91)90025-S.

[22] Friedman WE. Double fertilization in non-flowering seed plants and its relevance to the origin of flowering plants. Int. Rev. Cytol. 1992; 140:319-355. DOI:10.1016/S0074-7696(08)61101-8. 94.

[23] Floyd SK, Lerner VT, Friedman WE. A developmental and evolutionary analysis of embryology of *Platanus* (Platanaceae), a basal eudicot. Am. J. Bot. 1999; 86:1523-1537. https//doi.org/102307/2656790

[24] Bowe LM, Coat G, dePamphills CW. Phylogeny of seed plants based on all three genomic compartments: Extant gymnosperms are monophyletic and Gnetales' closes relative are conifers. Proc. Natl. Acad. Sci. USA. 2000; 97:4092-4097. DOI:10.1073/pnas.97.8.4092

[25] Friedman WE, Gifford EM. Development of the male gametophyte of Ginkgo biloba. In: *Hori T, editor*. *Ginkgo biloba* – a Global Treasure. New York, *Tokyo*: Springer-Verlag; 1997. *p. 29-49.*

[26] Friedman WE, Williams JH. Developmental evolution of the sexual process in ancient flowering plant lineages. Plant Cell. 2004. DOI: https://org/10.1105/tpc.017277

[27] Scutt CP. The origin of angiosperms. In: Nuno de la Rosa L, Müller G, editors. Evolutionary Developmental Biology. Springer, Cham. 2018. https://doi.org/10.1007/978-3-319-33038-9_60-1

[28] Porsch O. Versuch einer phylogenetischen Erklärung des Embryosackes und doppelter Befruchtung der Angiospermen. Jena: Gustav Fischer. Germany; 1907.

[29] Favre-Duchartre M. Interprétation de la reproduction sexuée des angiospermes a la lumiére d'organisation gametophytiqus et embryonnaires d'autre Archegoniates. Phytomorphology. 1971; 21:353-361.

[30] Fagerlind F. Der tetrasporische Angiospermen Embryosack und dessen Bedeutung für das Verständnis des Entwicklungsmechanik und Phylogenie des Embryosackes. Ark. Bot. 1944; 31A:1-71.

[31] Fagerlind F. Strobilus und Blüte von *Gnetum* und die Möglichkeit, aus ihrer Struktur den Blütenbau der Angiospermen zu deuten. Ark. Bot. 1946; 33A:1-57.

[32] Romanov ID. Types of embryo sac development in angiosperm plants. In: Problems of Embryology. Kyiv: Naukova Dumka. 1971. p. 72-113. (In Russian)

[33] Takhtajan AL. Foundations of Evolutionary Morphology of Angiosperms. Moscow-Leningrad: Nauka; 1964. 236 p. (In Russian)

[34] Kollman J. Das Ueberwintern von europaischen Frosch- unci Tritonlarven und die Umwandlung des mexikanischen Axolotl. Naturforschenden Gesellschaft in Basel. Verhancllungen. 1884; 7: 387-398.

[35] McNamara, K.J. Heterochrony: the evolution of development. Evo.Edu. Outreach. 2012; 5: 203-218. https://doi.org/10.1007/s12052-012-0420-3

[36] Buendía-Monreal M., Gillmor C.S. The times they are A-Changin': Heterochrony in plant development and evolutiont. Front. Plant Sci. https://doi.org/10.3389/fpls.2018.01349

[37] Alberch P, Gould SJ, Oster GF, Wake DB. Size and shape in ontogeny

and phylogeny. Paleobiology. 1979;5:296-317. DOI: https://doi.org/10.1017/S0094837300006588

[38] McKinney ML, McNamara KJ, editors. Heterochrony: the Evolution of Ontogeny. 1st ed. New York: Plenum Press; 1991. 437 p. DOI:10.1007/978-1-4757-0773-1

[39] Stebbins GL. Comparative aspects of plant morphogenesis: a cellular, molecular, and evolutionary approach. Am. J. Bot. 1992;79:589-598.

[40] Williams JH. Novelties of the flowering plant pollen tube underlie diversification of a key life history stage. PNAS. 2008;105:11259-11263. https://doi.org/10.1073/pnas.0800036105

[41] Williams JH. *Amborella trichopoda* (Amborellaceae) and the evolutionary developmental origins of the angiosperm progamic phase. Am. J. Bot. 2009; 96:144-165. DOI: 10.3732/ajb.0800070

[42] Williams JH. The evolution of pollen germination timing in flowering plants: *Austrobaileya scandens* (Austrobaileyaceae), AoB PLANTS. 2012a. DOI: 10.1093/aobpla/pls010

[43] Williams J. Pollen tube growth rates and the diversification of flowering plant reproductive cycles. Int. J Plant Sci. 2012b; 173: 649-661. DOI:10.1086/665822

[44] Williams JH, Reese JB. Evolution of development of pollen performance. Curr. Topics Dev. Biol. 2019; 131: 299-336. DOI: 10.1016/bs.ctdb.2018.11.012

[45] Singh H. *Embryology of Gymnosperms.* Berlin: Gebruder Borntraeger; 1978. 302 p.

[46] Williams CG. Pollination and fertilization. In: Conifer Reproductive Biology. Dordrecht: Springer; 2009. p. 91-105. https://doi.org/10.1007/978-1-4020-9602-0_6

[47] Fernando DD, Quinn CR, Brenner ED, Owens JN. Male gametophyte development and evolution in extant gymnosperms. Int. J. Plant Dev. Biol. 2010; 4:47-63.

[48] Favre-Duchartre M. Time relations and sexual reproduction in *Cichorium* and other angiosperms as compared with archegoniates. Phytomorphology. 1979; 29:166-178.

[49] Franchi GG, Nepi M, Dafni A, Pacini E. Partially hydrated pollen: taxonomic distribution, ecological and evolutionary significance. Plant Syst. Evol. 2002; 234:211-227. https://www.jstor.org/stable/23645003

[50] Dafni A, Firmage D. Pollen viability and longevity: practical, ecological and evolutionary implications. Plant Syst. Evol. 2000; 222:113-132. https://doi.org/10.1007/BF00984098

[51] Poddubnaya-Arnoldi VA. Characteristics of Angiosperm Plant Families on Embryological Traits. Moscow: Nauka; 1982. 352 p. (In Russian)

[52] Oksiuk PF, Khudyak MI. New data about fertilization in wheat. Proc. USSR Acad. Sci. 1955;105: 835-838. (in Russian)

[53] Kordyum EL. Pollination–fertilization in some species of Ranunculaceae family. Ukr. Bot. J. 1960; 17: 66-75. (in Ukrainian)

[54] Kravets OA, Khvedynich OA. Polyspermy and fertilization of antipodals in *Lilium* (Tourn.) L. species. Cyt. Gen. 2008;42:174-178. DOI: 10.3103/S0095452708030043

[55] Friedman WE. Expression of the cell cycle in sperm of *Arabidopsis*: implications for understanding patterns

of gametogenesis and fertilization in plants and other eukaryotes. Development. 1999;126:1065-1075. https://doi.org/10.1242/dev.126.5.1065

[56] Schnarf K. Vergleichende Embryologie der Angiospermen. Berlin: Gebruder Borntraeger; 1931. 354 p.

[57] Mauritzon J. Contribution to the embryology of the orders Rosales and Myrtales. Lunde Univ. Arsskr. Avd. 2. 1939; 35:3-121.

[58] Maheshwari P. An Introduction to the Embryology of Angiosperms. New York: McGraw Hill; 1950. 453 p.

[59] Gasser CS, Skinner DJ. Development and evolution of the unique ovules of flowering plants. Curr. Top. Dev. Biol. 2019; 131: 373-399. DOI: 10.1016/bs.ctdb.2018.10.007

[60] Kordyum EL. Evolutionary Cytoembryology of Angiosperm Plants. Kiev: Naukova Dumka; 1978. 219 p. (in Russian)

[61] Davis GL. Systematic Embryology of the Angiosperms. New York: Wiley; 1966. 528 p.

[62] Gifford EM. The shoot apex in angiosperms. Bot. Rev. 1954; 20: 477-529. https://doi.org/10.1007/BF02957569

[63] Hughes RNA. Functional biology of clonal animals. The growth and form of modular organisms. In: Harper JL, Rosen R, White J, editors. Functional Biology Series. London: Chapman and Hall, 1989.

[64] Fagerström T, Briscoe DA, Sunnucks P. Evolution of mitotic cell lineages in multicellular organisms. Trends Ecol. Evol. 1998;13:117-120. DOI: 10.1016/S0169-5347(97)01314-1

[65] Pineda-Krch M, Fagerström T. On the potential for evolutionary change in meristematic cell lineages through intraorganismal selection. J. Evol. Biol. 1999; 12:681-688. https://doi.org/10.1046/j.1420-9101.1999.00066.x

[66] Marcotrigiano M. Herbivory could unlock mutations sequestered in stratified shoot apices of genetic mosaics. Am. J. Bot. 2000;87:355-361. https://doi.org/10.2307/2656631

[67] Klekowski EJ, Kazarinova-Fukshansky N, Mohr H. Shoot apical meristems and mutation: Stratified meristems and angiosperm evolution. Am. J. Bot. 1985;72:1788-1800. https://doi.org/10.1002/j.1537-2197.1985.tb08451.x

[68] Roze D, Michod RE. Mutation, multilevel selection, and the evolution of propagule size during the origin of multicellularity. Am. Nat. 2001;158: 638-654. DOI:0003-0147/2001/15806-0007

[69] Bell G J. The sexual nature of the eucaryote genome. Heredity. 1993;84:351 – 359. doi: 10.1093/oxfordjournals.jhered.a111356.

[70] Meyer KI. Morphogeny of Higher Plants. Moscow: Moscow Univ. Publ. House; 1958. 255 p.

[71] Warming E. Untersuchungen über PollenbindendePhyllome and Kaulome. Hanstein's Bot. Abhandl.1873; 2:1-90.

[72] Maheshwari PA. Critical review of the types of embryo sacs in angiosperms. New Phytol. 1937; 36, 359-417. https://doi.org/10.1111/j.1469-8137.1937.tb06921.x

[73] Haig D. Poles apart: monosporic, bisporic, and tetrasporic embryo sacs revisited. Front. Ecol. Evol. 2020; 8. doi:10.3389/fevo.2020.516640

[74] Lopes MA, Larkins BA. Endosperm origin, development, and function. Plant Cell. 1993; 5:1383-1399. DOI:10.1105/tpc.5.10

[75] Friedman WE. The evolution of double fertilization and endosperm: An "historical" perspective. Sex. Plant Reprod. 1998;11:6-16. DOI:10.1007/s004970050114

[76] Khudyak MI. Endosperm of Angiosperm Plants. Kyiv: Ukr. SSR Acad. Sci. Publ. House; 1963. 182 p. (In Russian)

[77] Baroux C, Spillane C, Grossniklaus U. Evolutionary origins of the endosperm in flowering plants. Genome Biol. DOI:10.1186/gb-2002-3-9-reviews1026

[78] Li J, Berger F. Endosperm: food for humankind and fodder for scientific discoveries. Plant Physiol. 2012; 195:290-305. DOI:10.1111/j.1469-8137.2012.04182.x

[79] Kordyum EL, Mosyakin SL. Endosperm of angiosperms and genomic imprinting. Life. 2020; 10:1-20. DOI: 10.3390/life10070104

[80] Friedman WE, Williams JH. Modularity of the angiosperm female gametophyte and its bearing on the early evolution of endosperm in flowering plants. Evolution. 2003; 57:216-230. DOI:10.1111/j.0014-3820.2003.tb00257. 110

[81] Rueffler C, Hermisson J and Wagner GP. Evolution of functional specialization and division of labor. PNAS. 2012;109:326-335. https://doi.org/10.1073/pnas.1110521109

[82] Davis GL. Embryological studies in the compositae. 1. Sporogenesis, gametogenesis and embryogeny in *Cotula australis* (Less) Hook. Austral. J. Bot. 1962; 10:1-12

[83] Pacini, E., Franchi, G. G., Hesse, M. The tapetum: Its form, function, and possible phylogeny in Embryophyta. Pl Syst Evol. 1985; 149:155-185. DOI:10.1007/bf00983304

[84] Kravets E. A., Sidorchuk Yu. V., Horyunova I. I., Plohovskaya S. H., Mursalimov S. R., Deineko E. V., Yemets A. I., Blume Ya. B. Intra- and intertissular cytomictic interactions in the microsporogenesis of mono- and dicotyledonous plants. Cytol. Gen. 2016; 50: 267-277. DOI: 10.3103/S0095452716050054

[85] Zhao D. Z., Wang, G. F., Speal, B., Ma, H. The Excess microsporocytes1 gene encodes a putative leucine-rich repeat receptor protein kinase that controls somatic and reproductive cell fates in the *Arabidopsis* anther. Genes Dev., 2002, 16: 2021-2031.

[86] Tan H., Liang W., Hu J., Zhang D. MTR1 encodes a secretory fasciclin glycoprotein required for male reproductive development in rice, *Dev.* Cell, 2012, vol. 22, pp. 1127-1137.

[87] Furness, C. A., Rudall, P. J. The tapetum in basal angiosperms: Early diversity. Int. J. Plant Sci. 2001; 162: 375-392. DOI:10.1086/319580

[88] Johri BM, Ambegaokar KB. Some unusual features in the embryology of angiosperms. Proc. Indian Acad. Sci. (Plant Sci). 1984; 93:413-427.

[89] Mikesell J. Anatomy of terminal haustoria in the ovule of plantain (*Plantago major* L.) with taxonomic comparison to other angiosperm taxa. Bot. Gaz. 1990;151:452-464. https://www.jstor.org stable/2995332

[90] Agarwal S. Embryology of *Quinchamalimn chilense* Lato: in Plant Embryology: A symposium. New Delhi: CSIR; 1962; p. 162-169.

[91] Agarwal S. Morphoiogical and embryological studies in the family Olacaceae I. *Olax* L. Phytomorphology. 1963;13:185-196.

[92] Kuijt J. Haustoria of phanerogamic parasites. Ann. Rev. Phytopathol.

1977;17:91-118. https://doi.org/10.1146/annurev.py.15.090177.000515

[93] Arnorr HJ. The Seed, Germination and Seedling of *Yucca*. Univ. Calif Publ Bot. 1962; 35:1-144

[94] Tomlinsopn PB, Metcalfe CR. Anatomy of the monocotyledons. II. Palmae. C R Metcalfe, editor. Oxford: Clarendon; 1961

[95] Mauseth JD. Plant Anatomy. Benjamin/Cummings. Menlo Park. Calif.; 1988. 560 p. ISBN 0-8053-4570-1

[96] Swamy BGL. The embryo of monocotyledons: A working hypothesis from a new approach. In: Plant Embryology. New Delhi; 1962. p.113-123.

[97] Swamy GL, Rao JD. The endosperm in *Opilia amentacea* Roxb. Phytomorphology. 1963;13:423-428.

[98] Tiagi B. Studies in the family Orobanchaceae II. A contribution to the embryology of Aeginetia indica. Linn. Bull. Torrey Bot. Club. 1952;79: 63-68. https://doi.org/10.2307/2482106

[99] Soueges R. Embryogenie des Papilionacees Development de l'embryon chez le *Phaseolus vulgaris* L. C. R. Acad. Sci. Paris. 1950; 231: 637-640.

[100] Nagl W. The *Phaseolus* suspensor and its polytene chromosomes. Z. Pflanzenernähr. Bodenkd. 1974; 73: 1-44

[101] Kapil RN, Tiwari SC. The Integumentary tapetum. Bot. Rev. 1978;44: 457-490. https://www.jstor.org/stable/4353941

[102] Shamrov II. Diversity and typification of ovules in flowering plants. Wulfenia. 2018; 25: 81-109. https://www.zobodat.at › Wulfenia_25_0081-0109

[103] Swamy BGL, Thirumalachar MJ. Homologies of the embryo sac of angiosperms. Ann. Bot. 1945; 9:179-182.

[104] Bouman F. The ovule. In: Johri BM, editor. Embryology of angiosperms. New York: Springer Verlag; 1984. p. 123-157.

[105] Endress PK. Angiosperm ovules: diversity, development, evolution. Ann Bot. 2011; 107:1465-1489. DOI:10.1093/aob/mcr120

[106] Friis EM, Crane PR, Pedersen KR The endothelium in seeds of early angiosperms. 2019; 224: 1419-1424. https://doi.org/10.1111/nph.16024

[107] Płachno BJ, Świątek P, Kozieradzka-Kiszkurno M, Szeląg Z, Stolarczyk P. Integument cell gelatinization – the fate of the integumentary cells in *Hieracium* and *Pilosella* (Asteraceae). Protoplasma. 2017; 254:2287-2294. DOI: 10.1007/s00709-015-0894-2

[108] Musiał K, Płachno BJ, Świątek P, Marciniuk J. Anatomy of ovary and ovule in dandelions (*Taraxacum*, Asteraceae). Protoplasma. 2013; 250:715-722. DOI: 10.1007/s00709-012-0455-x.

[109] Engell K, Petersen GB. Integumentary and endothelial cells of *Bellis perennis*. Bot Tidsskrift. 1977; 71:237-244. http://pascal-francis.inist.fr/vibad/index.php?action=getRecordDetail&idt=PASCAL7710444366

[110] Ghimire B, Heo K. Embryology of *Abeliophyllum* (Oleaceae) and its phylogenetic relationships. Nordic J. Bot. 2014; 32: 632-641. https://doi.org/10.1111/j.1756-1051.2013.00204.x

[111] Colombo L, Franken J, Van der Krol AR, Wittich PE, Dons HJ, Angenent GC. Downregulation of ovule-specific MADS box genes from *Petunia* results in maternally controlled

defects in seed development. Plant Cell. 1997; 9: 703– 715. doi: 10.1105/tpc.9.5.703

[112] Debeaujon I, Peeters AJM, Léon-Kloosterziel KM, Koornneef M. The TRANSPARENT TESTA12 gene of *Arabidopsis* encodes a multidrug secondary transporter-like protein required for flavonoid sequestration in vacuoles of the seed coat endothelium. Plant Cell. 2001; 13: 853– 871.doi: 10.1105/tpc.13.4.853.

[113] Plisko MA. An electron microscopic investigation of the characteristic features of megagametogenesis in *Calendula officinalis* L. Bot. Zhur. 1971a; 56: 582-597. (in Russian)

[114] Plisko MA. Ultrastructure of integumental tapetum in *Calendula officinalis* in the early period of female gametophyte development. Zytologiya. 1971b;13:1211-1218. (in Russian)

[115] Povilus RA, Losada JM, Friedman WE. Floral biology and ovule and seed ontogeny of *Nymphaea thermarum*, a water lily at the brink of extinction with potential as a model system for basal angiosperms. Ann. Bot. 2015; 115: 211-226. DOI: 10.1093/aob/mcu235

[116] Tobe H, Stuessy TF, Raven PH, Oginuma K. Embryology and karyomorphology of Lactoridaceae. Amer J Bot. 1993; 80: 933– 946. https://doi.org/10.1002/j.1537-2197.1993.tb15315.x

[117] Yamaura A. Karyologisch und embryologische Studien uber einige Bambusa-Arten. (Vorlaufige Mittellung). Bot. Mag. Tokyo. 1933;47:551-555.

[118] Bohdanowicz J, Turała-Szybowska K. Ultrastructure of endopolyploid antipodals in *Aconitum vulparia* Rchb.: II. Antipodals in the period of free nuclear endosperm. Protoplasma. 1987;140:13-21 DOI: 10.1007/BF0127325

[119] Modilevsky YaS, Oksiuk PF, Khudyak MI, Dsyubenko LK, Beylis-Virovaya RA. Cytoembryology of Main Cereals. Kyiv: Ukr. SSR Acad. Sci. Publ. House. Ukraine; 1958. 336 p. (in Russian)

[120] Nagl W. The polytenic antipodal cells *Scilla bifolia*: DNA replication pattern and possibility of nucleolar DNA amplification. Cytobiologie. 1976;14: 165-170

[121] Doronina TV, Chaban IA, Lazareva EM. Structural and functional features of the wheat embryo sac's antipodal cells during differentiation. Rus. J. Dev. Bio. 2019; 50:194-208. DOI: 10.1134/S1062360419040039

[122] Zhang WC, Yan WM, Lou CH. The structural changes during the degeneration process of antipodal complex and its function to endosperm formation in wheat caryopsis. Acta Biol. Cracov. Ser. Bot. 1988; 30: 457-462.

[123] Nanchahal A, Kapil RN. Role of antipodal cells in poppy. Caryologia - Firenze-.2014; 40:195-205. DOI: 10.1080/00087114.1987.10797823

[124] Diboll AG. Fine structural development of the megagametophyte of *Zea mays* following fertilization. Am. J. Bot. 1968; 55: 797-806. DOI: 10.2307/2440968

[125] Chettoor AM, Evans MMS. Correlation between a loss of auxin signaling and a loss of proliferation in maize antipodal cells. Front. Plant Sci. 2015; 6: 187. DOI: 10.3389/fpls.2015.00187

[126] Wittich PE, Vreugdenhil D. Localization of sucrose synthase activity in developing maize kernels by *in situ* enzyme histochemistry. J. Exp. Bot.

1998; 49:1163-1171. DOI:10.1093/jxb/49.324.1163

[127] Souèges R. Recherches sur l'embryogénie des Renonculacées (Suite). Bull. Soc. Bot. France. 1910; 57:102-197.

[128] Schnarf K. Embryologie der Angiospermen. Berlin, Gebruder Borntraeger; 1927-1929. 417 S.

[129] Ivanovskay EV. Functional morphology of polytene chromosomes of wheat antipodes. Zytologiya. 1973; 15:1445-1452. (in Russian)

[130] Takhtadjan A.L. System and Phylogeny of Flowering Plants. Moscow–Leningrad: Nauka; 1966 (in Russian)

[131] Xue T, Zheng X, Chen D. et al. A high-quality genome provides insights into the new taxonomic status and genomic characteristics of *Cladopus chinensis* (Podostemaceae). Hortic. Res. 2020;7. DOI: 10.1038/s41438-020-0269-5

[132] Nagendran CR, Arekal GD, Subramanyam K. Embryo sac studies in three indian species of *Polypleurum* (Podostemaceae). Plant Syst. Evol. 1977; 128: 215-226. https://www.jstor.org/stable/23642222

[133] Sikolia S, Ochora P. Female gametophyte in Tristichoideae (Podostemaceae): Re-investigation. J. Biol. Sci. 2008;8:1158-1165. DOI: 10.3923/jbs.2008.1158.1165

[134] Sicolia S, Onyango JC. Female gametopyte in two Kenyan species of *Inverso dicraea* (Podostomaceae) Res. J. Bot. 2009; 4: 29-39. DOI: 10.3923/rjb.2009.29.39

[135] Murguía G, Novelo RA, Thomas P. et al. Embryo sac development in *Vanroyenella plumosa*, Podostemaceae Aquatic Bot. 2002;73:201-210. DOI: 10.1016/S0304-3770(02)00025-6

[136] Williams JH, Friedman WE. The four-celled female gametophyte of *Illicium* (Illiciaceae; Austrobaileyales): implications for understanding the origin and early evolution of monocots, eumagnoliids, and eudicots. Am. J. Bot. 2004; 91:332-351. DOI: 10.3732/ajb.91.3.332

[137] Krishnan R, Khanduri P, Tandon R. Floral biology, pollination mechanism and embryo development in *Zeylanidium maheshwarii* (Podostemaceae). Vegetos. 2019;32: 216-222. DOI: 10.1007/s42535-019-00025-4

[138] Sikolia SF. A critical review of the female gametophyte in the Podostemaceae – Past, Present and Future. J. Res. Agriculture Animal Sci. 2017;4: 12-28.

[139] Bhatnagar SP, Johri BM. Embryology of Loranthaceae. In: Calder M, Bernhardt P, editors. The Biology of Mistletoes, 47-67. New York, London, Sydney: Academic Press; 1983

[140] Bhandari NN, Vohra SCA. Embryology and affinities of Viscaceae. In: Calder M, Bernhardt P, editors. The biology of mistletoes, 69-86. New York. London. Sydney: Academic Press; 1983

[141] Kuijt J. The Biology of Parasitic Flowering Plants. Berkeley: University of California Press. CA, USA; 1969.

[142] Joshi PC. Morphological and embryological studies in the family Santalaceae. 5. *Osyris wightiana* Wall. Phytomorphology. 1960; 10: 239-248.

[143] Bachelier JB, Friedman WE. Female gamete competition in an ancient angiosperm lineage. PNAS USA. 2011;108(30):12360-12365. DOI: 10.1073/pnas.1104697108

[144] Sastri PLN. Studies in Lauraceae. 3. Embryology of *Cassytha*. Bot. Gaz. 1962;123:197-206.

[145] Bhandari NN, Nanda K. Studies in Viscaceae. 1. Morphology and embryology of the indian dwarf mistletoe *Arceuthobium minutissimum*. Phytomorphology. 1968;18:435-450.

[146] Ross CM., Sumner MJ. Early Endosperm and embryo development of the dwarf mistletoe *Arceuthobium americanum* (Viscaceae) Int. J. Plant. Sci. 2005 166: 901-907. https://doi.org/10.1086/432897

[147] Gonzalez AM, Sato HA and Marazzi B. Embryology in *Helosis cayennensis* (Balanophoraceae): Structure of female flowers, Fruit, endosperm and embryo. Plants. 2019; 8: 74. DOI: 10.3390/plants8030074

[148] Terekhin ES. On modi, directions and evolutionary significance of reduction of biological structures. Bot. J. 1975; 60: 401-412. (in Russian)

[149] Wolf YuI, Koonin EV. Genome reduction as the dominant mode of evolution. Bioessays. 2013; 35:829-837. https://doi.org/10.1002/bies.201300037

[150] Friedman WE, Williams JH. Developmental evolution of the sexual process in ancient flowering plant lineages. Plant Cell. 2004;16:S119-S132. DOI: https://doi.org/10.1105/tpc.017277

[151] Friedman WE, Moore RC, Purugganan MD. The evolution of plant development. Am. J. Bot. 2004; 91: 1726-1741. https://doi.org/10.3732/ajb.91.10.1726

[152] Müller GB. Evo-devo: extending the evolutionary synthesis. Nat. Rev. Gen. 2007;8:943-949. DOI: 10.1038/nrg2219

[153] Rutishauser R, Moline P. Evo-devo and the search for homology ("sameness") in biological systems. Theory in Biosci. 2005;124:213-241

[154] Rutishauser R. EvoDevo: Past and future of continuum and process plant morphology. Philosophies. 2020; 5:41. DOI:10.3390/philosophies5040041 1-38

[155] Ambrose B.A., Ferrándiz C. Development and the evolution of plant form annual plant. Reviews book series, 45: The Evolution of Plant Form. 2018. https://doi.org/10.1002/9781119312994.apr0494

Chapter 3

Pollinators: Their Relevance in Conservation and Sustainable Agro-Ecosystem

Mukesh Nitharwal, Rashmi Rolania, Hanuman Singh Jatav, Kailash Chandra, Mudassar Ahmed Khan, Subhita Kumawat, Sanjay Kumar Attar and Shish Ram Dhaka

Abstract

Survival and reproduction of several wild plants and crops is mostly by insects pollinator, their recognition and importance have been increased in this climatic changing scenario, which affects the various aspects of their life cycle. According to an estimate, approximately 30,000 species of bees are known in entomology, and about 190 species of bees have been reported to be associated with pollination. There can be an established link between seed production and pollinator diversity, for the plants with a generalist pollination system. The increasing of human habitation affects insect pollinators in various ways, i.e. of habitat destruction, results in low availability of food sources, nesting, oviposition, resting, and mating sites. Pollinator availability restraints the geographical distribution of plant species, i.e. to develop an ecological niche of certain plant species. Failure of pollinator- plant interaction mutualism results in lower seed production and sometimes extirpation of plant population has been recorded. The declining pollinators' population strengthens existing plant-pollinator interaction or allows new pant pollinator interaction to form. Maintaining the commercial and wild pollinator populations and preventing future shortages of pollination services, therefore, is extremely significant.

Keywords: agro-ecosystem, ecology, plant-pollinator interaction, pollinator, pollination, sustainable

1. Introduction

Plant and pollinator interaction results in the pollination of various plants that are *self-incompatible*. These pollinating agents are important for the existence and reproduction of about 87.5 percent of wild species of plants [1, 2]. Pollinators are important to increase agriculture production and in ecosystem functions to increase plant genetic diversity [3, 4]. The total annual economic value of crop pollination around the world has been estimated at about €153 billion [5]. Klein et al. [4], surveyed that more than 87 of the world's leading food crops, representing 35 percent of global food production, depends upon animal pollinators, pollination by insect directly contributed about $20 billion and that of honey bee contribution was $14.6 billion in 2000 in the U.S [6]. The total economic value of insect pollination of

Chinese fruits and vegetables recorded about52.2 billion US dollars in 2008, which represented 25.5% of the total production value of the 44 crops produced in China [7]. The area among the pollinator-dependent crops has increased up to 300% during the last 50 years, both in temperate and tropical crops [8, 9]. Pollination is an essential prerequisite to seed and fruit development; it is a pivotal, keystone process and insects provide an important function in both natural and managed ecosystems [10, 11]. The mutual relationship has been found between insect pollinators and flowering plants, that in return nectar and pollen are major food rewards for pollinators.

Robbins et al. [12] reported that most accepted estimates indicate honeybees' account for at least 80% of all insect pollination. For decades the consequences of insect pollination have been documented in treaties by Free [13], McGregor [14], and Pesson and Louveaux [15]. In apiculture the most important species, *Apis mellifera* L. has been reported as the single most important pollinator [13, 14]. The pollinating potential of a single honeybee colony becomes evident when it is recognized that its bees make up to 4 million trips per year and that during each trip an average of 100 flowers are visited [16].

2. Pollinators diversity in agro-ecosystem

An enormous number of the world's insect diversity visits flowers for nutrition, but all are not efficient pollinators. Among crop foods, fibers, edible oils, medicines, and other valuable products, a significant production occurs due to the vital role of insects and other animal pollinators. In all types of ecosystems, bees are recognized as the most valuable pollinators, but their precise roles in pollination are not well documented. According to estimation, approximately 30,000 species of bees are known in entomology, and about 190 species of bees have been reported to be associated with pollination in North America. Some of the other noteworthy contributors in pollination are; alkali bee (*Nomia melanderi* Cockerell) found as solitary bees to pollinator in alfalfa and alfalfa leaf cutting bee (*Megachile rotundata* Fabr.) also play its role in the pollination of this crop [17–20]. Through artificial mud nesting tunnels, orchard bees (*Osmia* spp.) can be managed in the field for pollination, as orchard bees are far much better for apple pollination than honeybees [21]. Another, bumblebees (*Bombus* spp.) are pollinators of red clover (*Trifolium pratense* L.) and cranberries (*Vaccinium macrocarpon* Ait.), but difficult to manage in field conditions [22]. Carpenter bees (*Xylocopa* spp.) are the outstanding pollinators of vine crops, especially of passion fruit (*Passifloraedulis* Sims), giant granadilla (*P. quadrangularis* L.), kiwi-fruit (*Actinidia deliciosa* (A.Chev.) C.F. Liang & A.E. Ferguson), various gourds, and winged beans (*Psophocarpus tetragonolobus* (L.)) but there is not much available data on these pollinators.

The interdependency of plants and pollinators vary in their degree, some plants species depend primarily on a single species of pollinator, which in turn has restricted sources of pollen or nectar. One example of a closely dependent association is the interaction between plant Yucca (Agavaceae) and its pollinators, yucca moth (*Tegeticula* spp.), having a mutualism estimated to be more than 40 million years old [23]. Squash crops i.e. pumpkins and gourds are pollinated by specialized squash bees, *Peponapis* spp. and *Xenoglossa* spp., and are more manageable, they nest in underground burrows and become active at dawn, visiting cucurbit flowers until about midday when unisexual flowers close [24, 25]. There are many other pollinator native bees, which include sunflower bees (*Eumegachile pugnata* (Say) [26], blueberry bees (*Habropoda laboriosa* (Fabr.)) [27], and *Osmia ribifllori*s Michener (**Table 1**) [28–30]. Oil palm weevil, *Elaeidobius kamerunicus* as the most valuable in terms of the economic importance of this crop at the world trade level [31]. Non-biting midges, *Forcipomyia* spp., a specialty pollinator among the cocoa (*Theobromacacao* L.) crop, which breeds in rotting vegetation in its plantation [32] are a few of specific examples of pollination to produce the economic plants/crops that are in need of human beings.

Among the herbivorous insects, the interaction of butterflies and moths is found during both its larval and adult stages and the latter is involved in pollination (**Table 2**). Some of these are *Heliconius* butterfly [41, 42]; yucca moth [43], obligatory mutualisms are exceptional in order Lepidoptera of pollinators. Moths from the families Geometridae, Noctuidae, and Sphingidae are among the most studied moths to be known as pollinators, their pollinating activity takes place at night in many plants such as cacti, orchids, trees [44–47]. Many had also identified and reported thrips on flowers and they noticed that thrips feed on pollen so that they

Common name	Scientific name	Example of crop plants pollinated
Alkali bee	*Nomia melanderi*	Alfalfa, clover, mint
Blueberry bee	*Habropoda laboriosa*	Blueberry
Carpenter bee	*Xylocopa* spp.	Passion flower, eggplant, pepper
Digger bee	*Andrena*, *Colletes*, and *Melissods* spp.	Cotton, fruit trees
Alfalfa leafcutting bee	*Megachile rotundata*	Alfalfa
Blue orchard bee (mason bee)	*Osmia lignaria*	Almond, apple, sweet cherry
Squash and gourd bee	*Peponapis pruninosa* other *Peponapis* and *Xenoglossa* spp.	Squash, pumpkin, gourds
Sunflower bee	*Eumegachile pugnata*	Sunflower

Table 1.
Bees and specific plant fauna they visit and pollinate.

Plant	Pollinator insect	References
Beetles		
Asimina	*Euphoria sepulchralis*	[33, 34]
	Trichius spp.	
	Trichotinius lunulatus	
Calycanthus	*Calopterus truncates*	[35]
Calochortus	*Acanthoscelides* spp.	[36]
Linanthus	*Trichochrous* sp.	[37]
Butterflies		
Wild carnation, *Dianthus carthusianorum*	Butterfly species	[38]
Native plants of North America	Checkerspot butterfly, *Euphydryas editha bayensis*	[39]
Milkweed and other	Monarch butterfly, *Danaus plexippus*	[40]

Table 2.
Beetles, butterflies, and moths with specific plant fauna they visit and pollinate.

can be effective pollinators or minor or secondary pollinators of a wide variety of plants in agro-ecosystem or nature [48–52].

3. Pollinators and plant interaction

In an ecosystem, the interaction between the organisms favors co-evolution and it gradually helps to evolve together for some betterment or for existence in nature. Those plants having a generalist pollination system, have a link between pollinator diversity and seed production can also be established [53]. Pollination biology (**Figure 1**) draws attention to both evolutionary and ecological approaches i.e. the link between pollinator behavior and plant mating patterns, generalization, and specialization in a pollination system [24, 54, 55].

There are many reasons for which pollinators visit flowers, including feeding, pollen collection, warmth in some cases, oils and resins, as well as for shelter and mating rendezvous sites [56]. These plant and pollinator interactions as mutualisms sustain not only plant diversity, but also the diversity of an estimated 350,000 animal species, including insects, birds, and mammals [57–59]. Ratto et al. [60] reported an average 63% loss of fruit or seed production when vertebrate pollinators are excluded from the flowering plants' ecology they visit. These results often reported experimentally that selective exclusion of a single group of an effective pollinators from plant-pollinator interaction can result in the failure of plants to produce fruits or seeds.

Diversity of pollinators in habitat can compete for floral resources [61], the declining pollinators population strengthens existing plant-pollinator interaction or allow new plant-pollinator interaction to form [62, 63]. The diverse pollen feeding behavior by bee species is due to digestibility and nutritional content requirement fulfillment [64]. There are specialized flower plant-pollinator relationships like certain solitary bees species [65], reduction of these flower plants from habitat often results in the elimination of their specialist plant-pollinator populations. Viana et al. [66] evaluated more than 250 studies that showed the impact of landscape and pollinators interactions. The forage bees' ability to assess the nutritional value of pollen sources before establishing plant-pollinator interaction is valuable [67, 68]. A recent study by Armbruster [69] on pollination ecology mainly emphasizes three aspects, first ecological (pollination involving one or few kinds of plant and animals), second phenotypic (having specialized flowers or morphologies) and third is evolutionary (showing transitions towards increased specialization).

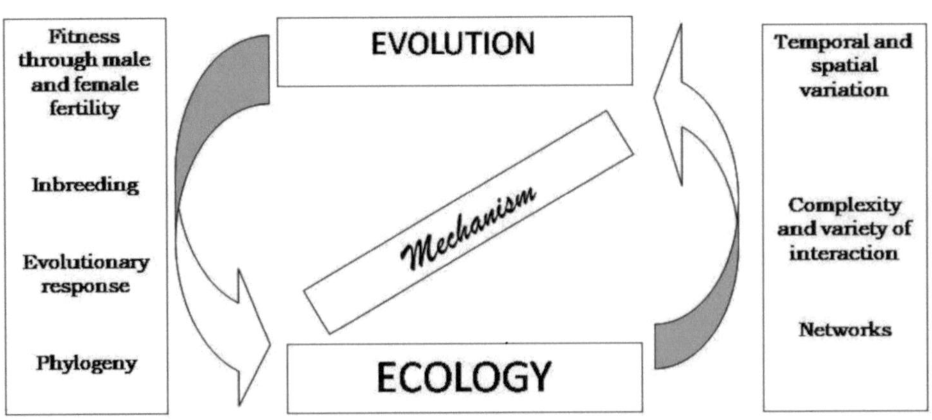

Figure 1.
Conceptual representation of the interplay between ecology and evolution in the study of plant-pollinator interactions.

4. Declining pollinator, a potential threat

It is difficult to determine as less surveys are organized to record whether pollinator species are declining around the world. If we study the literature many explanations have been invoked to account for declines in pollinator population around the globe [70–74]. There are a few of these reasons such as exposure to pathogens, parasites, and pesticides; habitat fragmentation and loss; climate change; market forces; intra and inter specific competition with native and invasive species; and genetic alterations. Reduction in pollinator diversity or abundance may influence the amount and source of pollen deposited on the reproductive part of flower or stigma [75].

The western honeybee, *Apis mellifera*, colony losses are elucidated in many literatures with attack of different honey pests i.e. parasites, pathogens [76], mostly predominately identified parasitic mites *Varroa destructor* and *Acarapis woodi*, the pathogen *Paenibacillus larvae* (American foulbrood) and the invasive Africanized honey bee [77–82].

The application of pesticides, especially insecticides in crops, vegetable, and orchards to control pests, kills or weakens thousands of honey bee colonies and affect their foraging and nesting behaviors that prevent plant pollination [83–86]. The basic behind pesticides to kill or weaken the colony is the result of accidents, careless application, or failure to adhere to label recommendations and warnings. Some of the advance studies showed that transgenic crops developed to reduce the unintended effects of pesticides have shown that there are direct effects on non-target species, including some pollinators [87–89]. Transgenic crops that express transgene with varied expression levels have not been yet reported effects on honeybee [90].

There are degradations reported around the globe in habitat i.e. alternations, fragmentation, and loss cause major problems for populations of many organisms, and pollinator populations are also one of them [1, 91]. Insect pollinator's i.e. bees and others require nesting sites (suitable soil, dead wood, abandoned mouse nests, and burrows) and floral resources (nectar and pollen) to exist. These habitat resources are at extinction through the disruption caused by row crop agriculture, grazing, and fragmentation of habitat into patches, which are small enough for the survival of diverse communities of pollinators [92]. Some other reported causes of decline in pollinators' population are monoculture, the lower density of weed flora, declining pastures, loss of flower-rich grasslands, and overgrazing can disrupt the nesting of bees [93–96].

Industrial development around the global, regional and local climate changes can alter or disrupt plant-pollinator relationships. Many studies and reports show the climate change forecast is shifted in temperature [97] and precipitation, concentrations of carbon dioxide [98] and ozone, and ultraviolet levels [99, 100] effects pollinators in many ways. There is evidence that the latitudinal and altitudinal ranges of some plants and pollinators have changed in the past 30 years, presumably in response to global warming [101–103].

5. Management and restoration of pollinators

Information on the status of most of the pollinators is incomplete around the world, and it is in a natal stage in developing countries [90]. Much can be done to maintain commercial and wild pollinator populations and to prevent future shortages of pollination services. Indigenous communities have an important role in the conservation of habitats through customary laws/rules, these areas are important

S. no.	Pollinator group	Resource function	Resource
1.	Honey bees, bumble bees	Nesting, roosting sites, or substrates	Cavities (underground, hollow trees)
2.	Nonsocial bees, wasps	Nesting sites or substrates	Bare ground, vertical cliffs or ditch bank, adobe walls
3.	Bumble bees	Nesting sites	Rodent, mouse nests
4.	Flies	Adult food	Pollen, nectare
5.	Leafcutter bees, mason bees	Building materials	Leaves cut into pieces or masticated
6.	Orchid bees	Pheromones	Essential oils, such as monoterpenoids collected by males
7.	Ants	Adult, larval food	Nectar, honeydew, insect prey

Table 3.
Pollinators and resources requirements.

biodiversity refuges providing valuable ecosystem services including pollination, which improved crop pollination in adjacent farming landscape [104–107]. An agri-environment scheme, on farmlands, has been proactively practiced in European Union countries through incentives to support biodiversity [108]. The US Farm bill (2008) had made specific economic provisions for pollinator conservation when it was further ratified in the 2014 Farm bill. The potential of conserving non-cropped land as a model in agro-ecosystem can be proved vital, through these agri- environment scheme models of conservation pollinators in the agriculture landscape can go a long way to inoculate pollinators naturally [109]. Mostly, pollinators are transported over long distances for the purpose of pollination [110]. They are also transported outside of their natural distribution range (e.g. African honey bees into Brazil, European bumble bees into Australia, Asia, and South America) [111].

Best management practices (BMP), similar to Good Agricultural Practice (GAP) should be promoted by the FAO in apiaries that need to be developed that respect local differences in beekeeping and hive management at the country level. There should be non-compulsive suggestions have been put forward overall lacking international harmonization [112]. In this perspective resistant stock of bees against parasitic mites is to be developed, identify the locally adapted stock of bees, instrumental insemination in bees, selection and managing miticide resistance in bees, etc.

Through collective approaches either for native and introduced bee species, whether solitary or social, requires the correct balance of water, flora hosts that offer sufficient pollen and nectar of the correct types [113], nest building materials (leaves, resin, sap, gums, floral oils, essential oils, bark, plant trichomes, old mouse nests, snail shells, mud, sand, pebbles), and nesting substrates [114] to survive as adults and rear their larval broods (**Table 3**) [115].

6. Limitations that restrict the pollinator-plant interaction

Deforestation and habitat changes have shown an adverse effect on insect pollinators, seed predators, decomposers, and parasitoids, which are highly susceptible to these changes. The success of plant reproduction may be sensitive to the loss of pollinators [74, 116, 117]. Some the evidence suggests that pollinator populations are declining worldwide [74]. These changes result in the destruction of food sources, nesting, oviposition, resting, and mating sites [95]. The increase in population pressure and urbanization of wild and agricultural lands has disrupted the habitat of wild pollinators viz., moths [118], butterflies [119], and bees [70, 71, 120, 121] and managed pollinators experienced sudden colony losses [122]. With the increased demand for food crops and higher productivity by use of either plant production or plant protection chemicals has killed the pollinators directly, and eradicated alternative pollen sources from their natural forage species [123, 124]. The pollination host range of honeybees is wide, but they do not pollinate all types of the crop with equal efficiency, are not active under all climatic conditions [125]. Whereas, some of the bees have the ability to pollinate some crops at a higher level of efficiency, with their lower population densities, and with greater independence of climatic conditions [21, 126].

Database of wild pollinator populations and communities is one of severe lack of long term planning and evaluation of their valuation in much of the world, especially for invertebrate pollinators [127–129]. Such as European red List of bees, 57% of the European bee species were categorized as "data deficient"; butterflies and moths from parts of Africa that are described at threatened status also reported in the literature [130–132].

Emerging risks such as diseases, invasive alien species, pathogens, etc., threaten the pollinators, there should be phytosanitary and preventive measures that could be accompanied for the effective response to these emerging risks. Few regions in the world (parts of Australia, Seychelles), that are not affected by the ectoparasitic mite, *Varroa destructor*, the most detrimental honey bee pest [82, 133, 134]. These mites also act as a vector for a number of bee viruses, which might spill over to wild bee species [135].

7. Conclusion and future prospect of pollinators

Sustainable agriculture requires reliable pollinators, but a shortage of pollinators could not be strongly evidenced for food crisis or pollination crisis. Long-term data deficient on the pollinators' population should be noted and there is no evidence of their decline over time, neither there is a framed definition to label pollinator crisis universally on that frame. The honeybee is a valuable pollinator to perform an important pollination function in the ecosystem. The decline in a number of managed pollinators in the system is due to some of the reasons such as introduced parasites and pathogens. There is a need for time to be compatible with and comprehensive management strategy of crop pollination for sustainable agriculture. Pollinators require to be managed through augmentation or conservation as needed to study their biology and ecology. Several studies that show the declining bee population poses a threat to global food security. Nesting habitat must be provided whether as a soil bed of a more or less special nature, or as stumps of trees and logs, or as rodent burrows for bumbles. Conservation of native pollinator habitat can be enhanced by changes in land use management strategies viz., non-cultivated patches of ground, setting up parks or protected areas for wildlife, flora, and fauna both at public and private areas. There should be a policy for arable, non-arable, and along with the roadside land that could facilitate the planting of wild plant flora which encourages pollinator populations. There should be judicious and timely use of pesticides that should ensure the protection of pollinators. There should be a crop pollination plan for all pollinator-dependent crops that must be included in the national or state crop production strategies. Farmer's awareness camps should be organized in the rural areas about crop pollination and the role of pollinators should be described, so that there may be a change in plant protection chemicals patterns.

Author details

Mukesh Nitharwal[1*], Rashmi Rolania[2], Hanuman Singh Jatav[1], Kailash Chandra[1], Mudassar Ahmed Khan[1], Subhita Kumawat[1], Sanjay Kumar Attar[1] and Shish Ram Dhaka[1]

1 S.K.N. Agriculture University, Jobner, India

2 University of Rajasthan, Jaipur, India

*Address all correspondence to: mnitharwal14@gamil.com

References

[1] Kearns CA, Inouye DW. Pollinators, flowering plants, and conservation biology. Bioscience. 1997;**47**:297-307

[2] Ollerton J et al. How many flowering plants are pollinated by animals? Oikos. 2011;**120**:321-326

[3] Cox PA, Elmqvist T. Pollinator extinction in the Pacific Islands. Conservation Biology. 2000;**14**:1237-1239

[4] Klein A-M, Vaissière B, Cane JH, Steffan-Dewenter I, Cunningham SA, Kremen C, et al. Importance of pollinators in changing landscapes for world crops. Proceedings of the Royal Society B. 2007;**274**:303-313

[5] Gallai N, Salles JM, Settele J, Vaissiere BE. Economic valuation of the vulnerability of world agriculture confronted with pollinator decline. Ecological Economics. 2009;**68**:810-821

[6] Barfield A, Bergstrom J, Ferreira S. An Economic Valuation of Pollination Services in Georgia. Birmingham: Southern Agricultural Economics Association; 2012

[7] Jian D, Chen WF. Economic value of insect pollination for fruits and vegetables in China. Acta Entomologica Sinica. 2011;**54**(4):443-450

[8] Aizen MA, Morales CL, Morales JM. Invasive mutualists erode native pollination Webs. PLoS Biology. 2008;**6**(2):e31. DOI: 10.1371/journal.pbio.0060031

[9] Aizen MA, Harder LD. The Global stock of domesticated honey bees is growing slower than agricultural demand for pollination. Current Biology. 2009;**19**:1-4

[10] Kevan PG, Baker HG. Insects as flower visitors and pollinators. Annual Review of Entomology. 1983;**28**:407-453

[11] Losey JE, Vaughan M. The economic value of ecological services provided by insects. Bioscience. 2006;**56**:311-323

[12] Robbins CS, Sauer JR, Greenberg RS, Droege S. Population declines in North American birds that migrate to the neotropics. Proceedings of the National Academy of Sciences USA. 1989;**86**(19):7658-7662

[13] Free JB. Insect Pollination of Crops. London: Academic Press; 1970. pp. 544

[14] McGregor SE. Insect pollination of cultivated crop plants. USDA Handbook 496. Washington: US Department of Agriculture, Agricultural Research Services; 1976. pp. 411

[15] Pesson P, Louveaux J, editors. Pollination et Productions Vegetables. Paris: INRA; 1984

[16] Free JB. Insect Pollination of Crops. 2nd ed. London, UK: Academic Press; 1993

[17] Cane JH. Alkali Bee. In: Shimanuki H, Flottum K, Harman A, editors. ABC & XYZ of Bee Culture. 41st ed. A.I. Root Company: Medina, OH; 2007. pp. 11-12

[18] Elfattah MSA. Sustainable use of leafcutting bee hives for alfalfa pollination. Munis Entomology & Zoology. 2010;**5**(2):807-811

[19] Mader E, Spivak M, Evans E. Managing Alternative Pollinators: A Handbook for Beekeepers, Growers, and Conservationists. Beltsville: Sustainable Agriculture Network; 2010. pp. 167

[20] Richards KW. Alfalfa leafcutter bee management in Western Canada. Ottawa: Agriculture Canada; 1984. pp. 51

[21] Torchio PF. Use of non-honey bee species as pollinators of crops. Proceedings of the Entomological Society of Ontario. 1987;**118**:111-124

[22] Plowright RC, Laverty TM. Bumble bees and crop pollination in Ontario. Proceedings of the Entomological Society of Ontario. 1987;**118**:155-160

[23] Pellmyr O. Yuccas, yucca moths, and coevolution: A review. Annals of the Missouri Botanical Garden. 2003; **90**:35-55

[24] Kevan PG. Pollination by animals and angiosperm biosystematics. In: Grant WF, editor. Plant Biosystematics. Toronto: Academic Press; 1984. pp. 271-292

[25] Tepedino VJ. The pollination efficiency of the squash bee (*Peponapispruinosa*) and the honey bee (*Apismellifera*) on summer squash (*Cucurbitapepo*). Journal of the Kansas Entomological Society. 1981;**54**:359-377

[26] Parker FD, Frohlich DR. Studies on management of the sunflower leafcutter bee *Eumegachilepugnata* (Say) (Hymentoptera: Megachilidae). Journal of Apicultural Research. 1985; **24**:125-131

[27] Sanchez D. Ground nesting bee best for blueberries. Agricultural Research. 1989;**37**(6):16-17

[28] Bosch J, Kemp W. How to Manage the Blue Orchard Bee as an Orchard Pollinator. Beltsville: Sustainable Agriculture Network; 2001. 88 pp. Available from: www.sare.org/publications/bee/blue_orchard_bee.pdf

[29] Torchio PF. Diversification of pollination strategies for U.S. crops. Environmental Entomology. 1990; **19**(6):1649-1656

[30] Torchio PF. *Osmia ribifloris* a native bee species developed as a commercially managed pollinator of highbush blueberry. Journal of the Kansas Entomological Society. 1990;**63**:427-436

[31] Syed RA. Studies on oil palm pollination by insects. Bulletin of Entomological Research. 1978;**69**:213-224

[32] Pouvreau A. Pollinisation du cacotier, du papayer, ettucafeier. In: Pesson P, Louveaux J, editors. Pollination et Productions Vegetales. Paris: INRA; 1984. pp. 195-208

[33] Kral R. A revision of *Asimina* and *Deeringothamus* (Annonaceae). Brittonia. 1960;**12**:233-278

[34] Norman EM, Clayton D. Reproductive biology of two Florida pawpaws; *Asmina ovata* and *A pygmaea*. Bulletin of the Torrey Botanical Club. 1986;**113**:16-22

[35] Grant V. The pollination of *Calycanthusoccidentalis*. American Journal of Botany. 1950;**37**:294-297

[36] Dilley JD. The radiation of *Calochortus*: Generalist flowers moving through a mosaic of potential pollinators. Oikos. 2000;**89**:209-222

[37] Grant KA, Grant V. Flower Pollination in the Phlox Family. New York: Columbia University Press; 1965

[38] Bloch D, Werdenberg N, Erhardt A. Pollination crisis in the butterfly pollinated, wild carnation *Dianthus carthusianorum*? New Phytologist. 2006;**169**:699-706

[39] Ehrlich PR, Hanski I, editors. On the Wings of Checkerspots: A Model System for Population Biology. New York: Oxford University Press; 2004

[40] Halpern S. Four Wings and a Prayer: Caught in the Mystery of the Monarch Butterfly. New York: Pantheon; 2001

[41] Boggs CL, Smiley JT, Gilbert LE. Patterns of pollen exploitation by Heliconius butterflies. Oecologia. 1981;**48**:284-289

[42] Estrada C, Jiggins CD. Patterns of pollen feeding and habitat preference

among *Heliconius* species. Ecological Entomology. 2002;**27**:448-456

[43] Pellmyr O, Thompson JN, Brown JM, Harrison RG. Evolution of pollination and mutualism in the yucca moth lineage. American Naturalist. 1996;**148**:827-847

[44] Gregory D. Hawkmoth pollination in the genus *Oenothera*. Aliso. 1963-1964;**5**:357-419

[45] Lin S, Bernardello G. Flower structure and reproductive biology in *Aspidospermaquebracho-blanco* (Apocynaceae), a tree pollinated by deceit. International Journal of Plant Sciences. 1999;**160**:869-878

[46] Little KJ, Dieringer G, Romano M. Pollination ecology, genetic diversity and selection on nectar spur length in *Platantheralacera* (Orchidaceae). Plant Species Biology. 2005;**20**:183-190

[47] Raguso RA, Henzel C, Buchmann SL, Nabhan GP. Trumpet flowers of the Sonoran Desert: Floral biology of *Peniocereus*cacti and Sacred Datura. International Journal of Plant Sciences. 2003;**164**:877-892

[48] Ananthakrishnan TN. The role of thrips in pollination. Current Science (Bangalore). 1993;**65**(3):262-264

[49] Kirk WDJ. Thrips and pollination biology. In: Ananthakrishnan TN, Raman A, editors. Dynamics of Insect-Plant Interaction. New Delhi: Oxford and IBH; 1988. pp. 129-135. [Chapter 10]

[50] Lewis T. Thrips: Their Biology, Ecology and Economic Importance. London and New York: Academic Press; 1973

[51] Terry I. Host selection, communication, and reproductive behaviour. In: Lewis T, editor. Thrips as Crop Pests. London: CAB International; 1997. pp. 64-118

[52] Terry I. Thrips and weevils as dual, specialist pollinators of the Australian cycad *Macrozamiacommunis* (Zamiaceae). International Journal of Plant Sciences. 2001;**162**:1293-1305

[53] Albrecht M, Schmid B, Hautier Y, Mueller CB. Diverse pollinator communities enhance plant reproductive success. Proceedings of the Royal Society B-Biological Sciences. 2012;**279**:4845-4852

[54] Davies TJ, Barraclough TG, Chase MW, Soltis PS, Soltis DE, Savolainen V. Darwin's abominable mystery: insights from a supertree of the angiosperms. Proceedings of the National Academy of Sciences, USA. 2004;**101**(7):1904-1909

[55] Labandeira CC, Dilcher DL, Davis DR, Wagner DL. Ninety-seven million years of angiosperm-insect association: Paleobiological insights into the meaning of coevolution. Proceedings of the National Academy of Sciences USA. 1994;**91**:12278-12282

[56] Simpson BB, Neff JL. Evolution and diversity of floral rewards. In: Jones CE, Little RJ, editors. Handbook of Experimental Pollination Biology. New York: Van Nostrand Reinhold Co.; 1983. pp. 142-159

[57] Bond WJ. Do mutualisms matter? Assessing the impact of pollinator and disperser disruption on plant extinction. Philosophical Transactions of the Royal Society of London Series B—Biological Sciences. 1994;**344**:83-90

[58] Ollerton J. Pollinator diversity: Distribution, ecological function, and conservation. In: Futuyma DJ, editor. Annual Review of Ecology, Evolution, and Systematics. Vol. 48. 2017. pp. 353-376

[59] Wilcock C, Neiland R. Pollination failure in plants: Why it happens and when it matters. Trends in Plant Science. 2002;7:270-277

[60] Ratto F, Simmons BI, Spake R, Zamora-Gutierrez V, MacDonald MA, Merriman JC, et al. Global importance of vertebrate pollinators for plant reproductive success: A meta-analysis. Frontiers in Ecology and the Environment. 2018;**16**(2):82-90

[61] Fort H. Quantitative predictions of pollinators' abundances from qualitative data on their interactions with plants and evidences of emergent neutrality. Oikos. 2014;**123**:1469-1478

[62] Kaiser-Bunbury CN, Muff S, Memmott J, Muller CB, Caflisch A. The robustness of pollination networks to the loss of species and interactions: A quantitative approach incorporating pollinator behaviour. Ecology Letters. 2010;**13**:442-452

[63] Memmott J, Craze PG, Waser NM, Price MV. Global warming and the disruption of plant-pollinator interactions. Ecology Letters. 2007;**10**:710-717

[64] Roulston TH, Cane JH. Pollen nutritional content and digestibility for animals. Plant Systematics and Evolution. 2000;**222**:187-209

[65] Muller A, Kuhlmann M. Pollen hosts of western palaearctic bees of the genus *Colletes* (Hymenoptera: Colletidae): The Asteraceae paradox. Biological Journal of the Linnean Society. 2008;**95**:719-733

[66] Viana BF, Boscolo D, Neto EM, Lopes LE, Lopes AV, Ferreira PA, et al. How well do we understand landscape effects on pollinators and pollination services? Journal of Pollination Ecology. 2012;**7**:31-41

[67] Nicholls E, Hempel de Ibarra N. Assessment of pollen rewards by foraging bees. Functional Ecology. 2017;**31**:76-87

[68] Ruedenauer FA, Spaethe J, Leonhardt SD. How to know which food is good for you: Bumblebees use taste to discriminate between different concentrations of food differing in nutrient content. Journal of Experimental Biology. 2015;**218**:2233-2240

[69] Armbruster WS. The specialization continuum in pollination systems: Diversity of concepts and implications for ecology, evolution and conservation. Functional Ecology. 2017;**31**:88-100

[70] Cameron SA, Lozier JD, Strange JP, Koch JB, Cordes N, Solter LF, et al. Patterns of widespread decline in North American bumble bees. Proceedings of the National Academy of Sciences of the United States of America. 2011;**108**:662-667

[71] Cameron SA et al. Patterns of widespread decline in North America bumble bee. Proceedings of the National Academy of Sciences of the United States of America. 2011;**108**:662-667

[72] Goulson D, Lye GC, Darvill B. Decline and conservation of bumble bees. Annual Review of Entomology. 2009;**53**:191-210

[73] Kerr JT, Pindar A, Galpern P, Potts SG, Roberts SM, Rasmont P, et al. Climate change impacts on bumblebees converge across continents. Science. 2015;**349**:177-180

[74] Potts SG, Biesmeijer JC, Kremen C, Neumann P, Schweiger O, Kunin WE. Global pollinator declines: Trends, impacts and drivers. Trends in Ecology and Evolution. 2010;**25**:345-353

[75] Aizen MA, Harder LD. Expanding the limits of the pollen limitation concept: Effects of pollen quantity and quality. Ecology. 2007;**88**:271-281

[76] Morse RA, Flottum K. Honey Bee Pests, Predators, and Diseases. 3rd ed. Medina, Ohio: The A. I. Root Company; 1997. 718 pp

[77] Barrett DP. History of American foulbrood in Michigan: Period

1927-1954. Gleanings in Bee Culture. 1955;**83**:460-461

[78] Evans JD. Diverse origins of tetracycline resistance in the honey bee bacterial pathogen Paenibacillus larvae. Journal of Invertebrate Pathology. 2003;**83**:46-50

[79] Gary NE Jr, Page RE, Morse RA, Henderson CE, Nasr ME, Lorenzen K. Comparative resistance of honey bees (Apismellifera L.) from Great Britain and United States to infestation by tracheal mites (Acarapiswoodi). American Bee Journal. 1990; **130**:667-669

[80] Kevan PG, Greco CF, Belaoussoff S. Log-normality of biodiversity and abundance in diagnosis and measuring of ecosystemic health: Pesticide stress on pollinators on blueberry heaths. Journal of Applied Ecology. 1997;**34**:1122-1136

[81] Nasr ME, Otis GW, Scott-Dupree CD. Resistance to Acarapiswoodi by honey bees (Hymenoptera: Apidae): Divergent selection and evaluation of selection progress. Journal of Economic Entomology. 2001;**94**(2):332-338

[82] Oldroyd BP, Wongsiri S. Asian Honey Bees: Biology, Conservation, and Human Interactions. Cambridge, Mass: Harvard University Press; 2006

[83] Decourtye A, Devillers J, Genecque E, Le Menach K, Budzinski H, Cluzeau S, et al. Comparative sublethal toxicity of nine pesticides on olfactory performances of the honeybee *Apismellifera*. Pesticide Biochemistry and Physiology. 2004;**78**:83-92

[84] Desneaux N, Decourtye A, Delpuech J. The sublethal effects of pesticides on beneficial arthropods. Annual Review of Entomology. 2007;**52**:81-106

[85] Johansen CA, Mayer DF. Pollinator Protection. A Bee and Pesticide Handbook. Cheshire, Conn: Wicwas Press; 1990

[86] Thompson HM. Behavioural effects of pesticides use in bees—Their potential for use in risk assessment. Ecotoxicology. 2003;**12**:317-330

[87] Losey JE, Rayor LS, Carter ME. Transgenic pollen harms monarch larvae. Nature. 1999;**399**:214

[88] Malone LA, Pham-Delègue MH. Effects of transgene products on honey bees (*Apismellifera*) and bumblebees (*Bombus*sp.). Apidologie. 2001;**32**:287-304

[89] O'Callaghan M, Glare TR, Burgess EPJ, Malone LA. Effects on plants genetically modified for insect resistance on non-target organisms. Annual Review of Entomology. 2005;**50**:271-292

[90] National Research Council (NRC). Status of Pollinators in North America. Washington, DC: The National Academies Press; 2007

[91] Kevan PG. Pollination: Plinth, pedestal, and pillar for terrestrial productivity. The why, how, and where of pollination protection, conservation, and promotion. In: Stubbs CS, Drummond FA, editors. Bees and Crop Pollination—Crisis, Crossroads, Conservation, Proceedings, Entomological Society of America. Lanham, Md.: Thomas Say Publications in Entomology; 2001. pp. 7-68

[92] Kearns CA, Inouye DW, Waser NM. Endangered mutualisms: The conservation of plant-pollinator interactions. Annual Review of Ecology and Systematics. 1998;**29**:83-112

[93] Gess FW, Gess SK. Effects of increasing land utilization on species representation and diversity of aculeate

wasps and bees in the semi-arid areas of southern Africa. In: LaSalle J, Gauld ID, editors. Hymenoptera and Biodiversity. Wallingford: CAB International; 1993. pp. 83-113

[94] Goulson D. Effects of introduced bees on native ecosystems. Annual Review of Ecology and Systematics. 2003;**34**:1-26

[95] Kevan PG. Pollinators as bioindicators of the state of the environment: Species, activity and diversity. Agriculture, Ecosystems and Environment. 1999;**74**:373-393

[96] Vinson SB, Frankie GW, Barthell J. Threats to the diversity of solitary bees in a neotropical dry forest in Central America. In: LaSalle J, Gauld ID, editors. Hymenoptera and Biodiversity. Oxon: CAB International; 1993. pp. 53-82

[97] Forister ML, Shapiro AM. Climatic trends and advancing spring flight of butterflies in lowland California. Global Change Biology. 2003;**9**:1130-1135

[98] Davis AR. Influence of elevated CO_2 and ultraviolet-B radiation levels on floral nectar production: A nectary-morphological perspective. Plant Systematics and Evolution. 2003;**238**:169-181

[99] Collins SA, Robinson GE, Conner JK. Foraging behavior of honey bees (Hymenoptera: Apidae) on *Brassica nigra*and *B. rapa*grown under simulated ambient and enhanced UV-B radiation. Annals of the Entomological Society of America. 1997;**90**(1):102-106

[100] Sampson BJ, Cane JH. Impact of enhanced ultraviolet-B radiation on flower, pollen, and nectar production. American Journal of Botany. 1999;**86**:108-114

[101] Crozier L. Winter warming facilitates range expansion: Cold tolerance of the butterfly *Atalopedescampetris*. Oecologia. 2003;**135**:648-656

[102] Walther GR. Plants in a warmer world. Perspectives in Plant Ecology, Evolution and Systematics. 2004;**6**:169-185

[103] Wilson R, Gutiérrez D, Gutiérrez J, Martínez D, Agudo R, Monserrat VJ. Changes to the elevational limits and extent of species ranges associated with climate change. Ecology Letters. 2005;**8**(11):1138

[104] Bhagwat S. Ecosystem services and sacred natural sites: Reconciling material and non-material values in nature conservation. Environmental Values. 2009;**18**:417-427

[105] Blicharska M, Mikusiński G, Godbole A, andSarnaik J. Safeguarding biodiversity and ecosystem services of sacred groves—Experiences from northern Western Ghats. International Journal of Biodiversity Science, Ecosystem Services & Management. 2013;**9**:339-346

[106] Bodin O, Tengo M, Norman M, Lundberg J, Elmqvis T. The value of small size: Loss of forest patches and ecological thresholds in Southern Madgascar. Ecological Applications. 2006;**16**:440-451

[107] ICCA. Bio-cultural Diversity Conserved by Indigenous Peoples & Local Communities—Examples & Analysis. 2012. Available from: http://www.cenesta.org/wp-content/uploads/2015/10/publication-ceesp-briefing-note-10-companion-en.pdf

[108] Batary P et al. Effect of conservation management on bees and insect-pollinated grassland plant communities in three European countries. Agriculture, Ecosystems and Environment. 2010;**136**:35-39

[109] Morandin LA, Winston ML. Pollinators provide economic incentive

to preserve natural land in agroecosystems. Agriculture Ecosystems and Environment. 2006;**116**:289-292

[110] Cavigli I, Daughenbaugh KF, Martin M, Lerch M, Banner K, Garcia E, et al. Pathogen prevalence and abundance in honey bee colonies involved in almond pollination. Apidologie. 2016;**47**:251-266

[111] Moritz RF, Härtel S, Neumann P. Global invasions of the western honeybee (Apismellifera) and the consequences for biodiversity. Ecoscience. 2005;**12**(3):289-301

[112] Ritter W. Good beekeeping practice—Knowledge in a Nutshell. Bees for Development Journal. 2013;**107**:3-5

[113] Roulston TH, Cane JH, Buchmann SL. What governs the protein content of pollen grains: pollinator preferences, pollen-pistil interactions, or phylogeny? Ecological Monographs. 2000;**70**:617-643

[114] Shepherd MD, Buchmann SL, Vaughan M, Black SH. Pollinator Conservation Handbook: A Guide to Understanding, Protecting, and Providing Habitat for Native Pollinator Insects. The Xerces Society in Association with The Bee Works. Portland: The Xerces Society; 2003. 145 pp

[115] Michener CD. Bees of the World. Baltimore and London: The Johns Hopkins University Press; 2000. 913 pp

[116] Burkle L, Marlin J, Knight T. Plant-pollinator interactions over 120 years: Loss of species, co-occurrence, and function. Science. 2013;**80**:1611-1615

[117] Gonzalez-Varo JP, Biesmeijer JC, Bommarco R, Potts SG, Schweiger O, Smith HG, et al. Combined effects of global change pressures on animal-mediated pollination. Trends in Ecology and Evolution. 2013;**28**:524-530

[118] Fox R. The decline of moths in Great Britain: A review of possible causes. Insect Conservation and Diversity. 2012;**6**:5-19

[119] Maes D, Van Dyck H. Butterfly diversity loss in Flanders (north Belgium): Europe's worst case scenario? Biological Conservation. 2001;**99**:263-276

[120] Biesmeijer JC et al. Parallel declines in pollinators and insect pollinated plants in Britain and the Netherlands. Science. 2006;**313**:351-354

[121] Kosior A et al. The decline of the bumble bees and cuckoo bees (Hymenoptera: Apidae: Bombini) of western and central Europe. Oryx. 2007;**41**:79-88

[122] van Engelsdorp D et al. A survey of honey bee colony losses in the U.S., fall 2007 to spring 2008. PLoS ONE. 2008;**3**:e4071

[123] Kevan PG. Pollination and environment conservation. Environmental Conservation. 1975;**2**:222-227

[124] Keven PG. Pollination and flower visiting insects and the management of beneficial and harmful insects and plants. In: Hussein MY, Ibrahim AG, editors. Biological Control in the Tropics. Serdang, Selangor, Malaysia: University Pertanian Malaysia; 1986. pp. 439-452

[125] Keven PG, editor. Alternative pollinators for Onario's crops. Proceedings of the Entomological Society of Ontario. 1987;**118**:109-170

[126] Boyle-Makowski RMD. The importance of native pollinators in cultivated orchards: Their abundance and activities in relation to weather conditions. Proceedings of the Entomological Society of Ontario. 1987;**118**:125-141

[127] Aguilar R, Ashworth L, Galetto L, Aizen MA. Plant reproductive susceptibility to habitat fragmentation: Review and synthesis through a meta-analysis. Ecology Letters. 2006;**9**:968-980

[128] Montero-Castano A, Vila M. Impact of landscape alteration and invasions on pollinators: A meta-analysis. Journal of Ecology. 2012;**100**:884-893

[129] Morales CL, Traveset A. A meta-analysis of impacts of alien vs. native plants on pollinator visitation and reproductive success of co-flowering native plants. Ecology Letters. 2009;**12**:716-728

[130] Edge DA, Mecenero S. Butterfly conservation in Southern Africa. Journal of Insect Conservation. 2015;**19**:325-339. DOI: 10.1007/s10841-015-9758-5

[131] Mecenero S, Ball JB, Edge DA, et al. Conservation Assessment of Butterflies of South Africa. Lesotho and Swaziland: Red List and Atlas; 2013

[132] Nieto A, Roberts SPM, Kemp J, Rasmont P, Kuhlmann M, Biesmeijer JC, et al. European Red List of Bees. Luxembourg: Publication Office of the European Union; 2014

[133] Anderson DL, Trueman JWH. *Varroajacobsoni* (Acari: Varroidae) is more than one species. Experimental and Applied Acarology. 2000;**24**:165-189

[134] Sammataro D, Gerson U, Needham G. Parasitic mites of honey bees: Life history, implications, and impact. Annual Review of Entomology. 2000;**45**:519-548

[135] McMahon DP, Fürst MA, Caspar J, Theodorou P, Brown MJF, Paxton RJ. A sting in the spit: Widespread cross-infection of multiple RNA viruses across wild and managed bees. The Journal of Animal Ecology. 2015;**84**:615-624

Chapter 4

Threat to *Citrus* in a Global Pollinator Decline Scenario: Current Understanding of Its Pollination Requirements and Future Directions

Subhankar Gurung and Arun Chettri

Abstract

Pollinators are vital for world biodiversity and their contribution to agricultural productivity is immense. Pollinators are globally declining with reports such as colony collapse being documented. *Citrus* exhibits a varying degree of pollination requirements due to its vast cultivars being developed all the time. The article intends to understand the breeding system of a few commercially important *Citrus* groups and discern its dependency on pollination services. The threat related to pollinator decline to the *Citrus* industry is measured not only by its reliance on pollinators but also the requirement of the consumers and manufacturers who mostly seek seedless varieties. Therefore, the threat can be tackled by developing high-quality seedless varieties where pollination requirement is absent. Although the importance of pollinators on several self-incompatible varieties cannot be negated, the impact of pollinator decline on its production will entirely depend upon the demand of the market.

Keywords: breeding system, pollinators, pollination, *Citrus*, fruit set

1. Introduction

Pollinators are insects or vectors that transfer the pollen from the male part of the flower, i.e., anther to the female part of the flower i.e., stigma enabling fertilization and seed production [1, 2]. Pollination is an ecosystem service that enhances crop production and helps in sustaining human life [1, 3]. Furthermore, many wild plants that provide calories and micronutrients to human diets require pollination as well [4]. Approximately, 90% of wild flowering plant species worldwide rely on insect vectors for pollen transmission. These plants are essential for ecosystems to function properly because they supply food, habitat, and other resources to a variety of different species [5]. One-third of the agricultural plants, including the cultivars, are dependent on animal pollinators for their reproduction and increased fruit set [6]. Bees play a pivotal role among pollinators by visiting more than 9% of the leading global crop plants [3]. Unfortunately, agricultural intensification has led to a loss of habitat of many insect pollinators and monoculture plantings have

IntechOpen

threatened insect biodiversity [7, 8]. A palpable concern for the global decline in pollinators can be sensed [9] which may affect fruit set or seed set [10] influencing its yield [11, 12]. The decline in pollinators would also result in a parallel decline in its associated plant species [9]. Insect pollination is therefore also known as an endangered ecosystem service [13–16]. The pollinator crisis threatens global and local food security and can destabilize human life [17]. Therefore, studies in crop pollination are upfront because of the perceived danger that it invites [16–18]. However, a detailed preview of pollination studies is incomplete or redundant [19]. Therefore, understanding the pollination needs of different cultivated plants is a prerequisite to understanding their dependency on animal pollinators [19].

In this regard, the pollination requirement of *Citrus* has been arising much curiosity. Domestication of *Citrus* sp. along with grafting has resulted in variation in the existing species in comparison to its wild progenitors [20, 21]. As such, the present cultivars exhibit tremendous variation in pollination requirements. *Citrus* was considered as a crop with little or no requirement for insect pollination [22]. Webber and Batchelor [23] emphasized, however, that no variety is ever static, and the possibility that its pollination requirements have changed cannot be ruled out. Further research over the period has proven Webber and Batchelor [23] to be correct. However, it has been found the pollination requirements of *Citrus* have been confusing and misleading. It has been reported that *Citrus* have complete or perfect flowers and generally pollinate themselves to produce fruit [24]. *Citrus* has been reported to exhibit varying breeding systems within species and cultivars of the same species [22, 25]. Pollinator reliance and breeding systems differ between and within *Citrus* species, and even between cultivars of the same species [25].

Several *Citrus* hybrid cultivars also exhibit sexual incompatibility which affects their yield [26]. Further, *Citrus* exhibit parthenocarpy and is an important trait that results in a seedless variety [27, 28]. Therefore, it would not be wrong to say that the pollination requirements for *Citrus* are quite diverse [22, 29]. The reason for such a varying breeding system can be attributed to its numerous varieties and the fact that more are being developed all the time. Moreover, each variety has its characteristics that determine the pollination mechanism in them [29]. For instance, some studies state that cross-pollination is not necessary for fruit sets [25, 30], whereas others showed that insect pollinators increased fruit production [31]. Most seeds are produced by asexual means in some cultivars. However, the benefits drawn from insect variation cannot be negated [22].

Given the complex breeding system of *Citrus*, it is difficult to predict the magnitude of the threat the dwindling pollinator number poses to the productivity of *Citrus*. As a step towards understanding the repercussion, we review the pollination requirements of some of the commercially important *Citrus* cultivars and highlight the danger that the global pollinator decline poses on such *Citrus* varieties.

2. Pollinators and its global decline

Honeybee (*Apis mellifera*) is the primary pollinating agent of *Citrus*. They have shown to be effective pollinators of an array of cultivated crops, oilseeds, forage crops, fiber crops, and cereal crops [32]. Evidence suggests that 59% of the domestic honeybee colonies were lost between 1947 and 2005 [33] and 25% were lost in Europe between 1985 and 2005 [34]. Considerable fear has been sensed about the future availability of honeybee pollination. Further, it was observed that specialists (diet/habitat) and sedentary bees and hoverflies tended to decline while mobile generalists tended to thrive [9]. Globally, *A. mellifera* is the predominantly managed pollinator to augment agricultural productivity [34] and ranks as the most frequent single species of pollinator for crops worldwide [35]. There has been increased mortality of managed *A. mellifera* colonies in some regions of the world [36, 37], which could spread to its wild population [38]. Considering the importance of *A. mellifera* in crop productivity, threats to its population could prove to be a disaster for crop reproduction and its productivity [1]. Further, reports suggest that the floral visits by *A. mellifera* were double that of all bumblebee species combined [38]. *A. mellifera* has also been observed to compete and displace native pollinators [39, 40] or compromise plant reproductive success [41]. Therefore, it is the competition between native and non-native pollinators along with various other factors such as pesticides, pathogens, parasites, and climate change that have been triggering the decline of pollinators worldwide [38].

3. Inflorescence of *Citrus* and its sexual process

A *Citrus* flower usually comprises both male and female sex organs in the same flower [26]. Each flower has radial symmetry and consists of 5–6 sepals and 5 petals (**Figure 1B**). The inflorescence is a cyme or raceme, rarely of solitary flowers. The flowers are usually bisexual and actinomorphic, hypogynous, rarely epigynous. It consists of 4–5 (2–3) sepals or lobes. The corolla is apopetalous or sympetalous with 4–5 (0, 2–3) imbricate or valvate petals or lobes. The stamens are 8–10–∞ usually diplostemonous in 2 (1–4) whorls.

The stigma and ovary are connected by a white style that forms a knob [42]. The pistil is in the center of each flower and is surrounded by a whorl of stamens which consists of long filaments with anthers at the top (**Figure 1A**). The anthers are longitudinal indehiscence and consist of yellow pollen that gets shed on the sticky

Figure 1.
*Flowers of mandarin (*Citrus reticulata*): (A and B) single flower, (C) fertilized ovary, (D) female sex organ (pistil); An—anther, Fi—filament, Pe—petals, Ov—oary, Ne—nectar, Sti—stigma, Sty—style.*
©Subhankar Gurung.

surface of the stigma [26]. *Citrus* pollen is heavy and sticky [26], and is not adapted for wind pollination [43]. *Citrus* flowers are either self-pollinated, cross-pollinated [43], or can produce fruit by parthenocarpy or nucellar embryony [44]. At the bottom of the ovary, there is a nectary gland. *Citrus* produces nectar copiously [22] and is secreted continuously for at least 48 hours after the flower opening [45].

3.1 Breeding system of mandarin and mandarin hybrid-complex

Several mandarin hybrids of fine quality have been evolved that have become commercially important varieties. For instance, crosses between clementine and tangelo (*Citrus reticulata* × *[Citrus paradisi* × *C. reticulata]*) have produced hybrids such as Robinson, Lee, Osceola, Page, and Nova [46]. King mandarin *Citrus nobilis* proved to be an exceptional parent when crossed with other mandarin types in producing well-flavored hybrids [47]. Tangelo is a cross between tangerine, or mandarin orange (*C. reticulata*) with either grapefruit (*Citrus paradisi*) or pummelo (*Citrus grandis*) [24]. Clementine (*Citrus clementina*) is a hybrid between Mediterranean *Citrus deliciosa* and sweet orange [48]. While the hybrids were being released, issues in its pollination were recognized where the fruit set in Clementine mandarin was fewer in flowers enclosed in self-pollination without bees than those flowers enclosed with bees [49]. Clementine mandarin was identified to be self-incompatible but showed improved fruit characteristics when cross-pollinated with March grapefruit and Balady orange [48]. Besides, it has also been reported that Clementine mandarin develops a seedless fruit if cross-pollination does not occur [50]. Clementine mandarin cultivars (*Citrus clementina*) exhibit facultative parthenocarpy and is pollination independent [50]. However, "Clemenules" exhibits weak parthenocarpic ability than 'Marisol' indicating a difference in the parthenocarpic ability of Clementine cultivars. Further, studies on "Lee", "Page", "Nova" and "Robinson" identified them to be self-incompatible [51]. 'Afourer' mandarin was also found to be self-incompatible and produced relatively high fruit set by facultative and autonomous parthenocarpy [52]. On the contrary, studies demonstrated that "Osceola" was not completely self-incompatible [53]. Further studies showed that "Hyuganatsu" mandarin was self-sterile but cross-fertile [54]. Several experiments demonstrated that "Satsuma" mandarin benefitted 6.3% [55] and 7–11% [56, 57] by bees. They also present obligate parthenocarpy [58], and thus the sexual process is not always required for successful fruit development. There was no significant difference in the final fruit set between parthenocarpic and fertilized Satsuma mandarin [50]. Several mandarin cultivars also benefit tremendously from compatible pollinizers for successful fruit development. For instance: grapefruit pollen is an efficient pollinizer in fertilizing "Clementine" mandarins. Further, field experiments demonstrated that pollens derived from "Lee" enhanced the fruit size of "Page" cultivars [59]. Minneola tangelo increased its yield when cross-pollinated with "Seminole" or "Lake" pollen which indicated that the Minneola fruit set can be enhanced by planting a suitable pollinizer [60].

Some of the accessions of mandarin such as "Kunenbo" are self-incompatible because of incompatible alleles originating from pummelo [61] while mandarins that originated in China and India are self-compatible [62].

3.2 Breeding system of sweet orange

Breeding system of oranges is complex because of the variation in its cultivars. *Citrus sinensis* var. Pera-Rio benefitted from the visitation of bees resulting in heavier, less acidic fruit with fewer seed sets [63]. Further, the fruit production was 35.30% greater in uncovered flowers, fruit weight was higher (180.2 g) than covered ones (168.5 g) [64]. However, bees did not affect the production of *C. sinensis* var. "Valencia" sweet oranges [65, 66]. Sanford [66] also mentions that 'Valencia' does not benefit much from pollination by bees. "Valencia" experienced lower pollen grain germination resulting in a lower seed set [50]. Defective ovule mutants encourage flawed pollen tube guidance [50], and therefore attributed low seed set in "Valencia" sweet orange to defects during female gametogenesis [67]. However, unlike the previous report, fertilization remains an important factor in its improved fruit set [50]. Similarly, ambiguity regarding the pollination of "washington navel" remained when studies demonstrated that cross-pollination might influence fruit set [68]. However, later studies reinstated that cross-pollination does not affect the yield of "washington navels" [69]. Atkins [70] stated that there is a possibility that cross-pollination by bees may cause them to retain more fruit [70]. There was a 31% increase in the fruit set, a 21% increase in fruit weight, 33% more juice, and 36% more seeds from fruits on trees visited by bees [71]. Further "washington navel" is male sterile and is incapable of sexual reproduction when used as a pollen parent [72]. Therefore, the likeliness of it requiring a pollinizer is high for a successful fruit set. In this context, it was observed that "washington navel" produced the most fruits when crossed with "Grusinian" orange [57]. However, the embryo sac of "washington navel" may also degenerate before pollen tubes penetrate through it.

3.3 Breeding system of pummelo

Most of the pummelo varieties are self-incompatible because pollen tubes cannot grow properly because of the genetic barrier to inbreeding [73]. Further, most pummelo trees are self-incompatible and should be inter-planted with other cultivars [74]. Shaddock (*Citrus maxima* L.) or (*Citrus grandis* L.) increased its fruit set %, fruit retention, fruit weight, fruit size when cross-pollinated with Balady orange pollen grains [48]. Further, Aala [74] suggested that bees are important for proper pollination and fruit set irrespective of a cultivar being self-fertile or self-sterile. *Citrus maxima* (burm) Merr cv. *nambangan* is also reported to be self-incompatible and is dependent on insect pollinators for fruit sets [42]. However, Hoang et al. [75] reported that pollen tube growth of *Citrus grandis* cv. "Phuc Trach", "Dien" and "Red Pummelo" were arrested in the style inhibiting fertilization. "Da Xanh" and "Nam Roi" pummelos did not require pollination and produced large-sized seedless fruit by vegetative parthenocarpy [75]. Therefore, it is an established fact that few cultivars of pummelo are highly parthenocarpic [27] and pollination is unnecessary for commercial fruit production.

3.4 Breeding system of grapefruit

The pollination requirements and breeding system of grapefruit are poorly understood [76]. Some studies suggest that cross-pollination is not a prerequisite for fruit set [25, 77], whereas others demonstrate that the transfer of cross-pollen mediated by insects increases fruit production [31]. For example, there are species such as "Red Blush" grapefruit cultivar that is highly parthenocarpic while "Marsh" grapefruit set four times as many fruits in open-pollinated flowers than selfed flowers [78]. "Star Ruby" grapefruit (*C. paradisi* Macf.) consists of non-functional pollen, non-functional ovules and produces more fruits when cross-pollinated [31]. Further, studies suggest that pollen transfer by insects and pollen tube development are important factors for fruit production in grapefruit [76, 79]. Chacoff and Aizen [76] also demonstrated that more cross-pollen tubes reached the ovary than self-pollen, indicating the importance of pollinators in successful fruit development.

3.5 Breeding system of lemons and limes

Lemons (*Citrus aurantiifolia* (Christm.) Swingle) is self-compatible and shows both autogamous and allogamous types of pollination [80]. On the contrary, "Tahiti" lime is strongly parthenocarpic and its dependency on pollination is feeble for increased fruit set [69]. Parthenocarpy in "Tahiti" lime is followed by chromosomal irregularity during meiosis, reflecting the triploid origin of the lines [81]. Likewise, parthenocarpy is high on thornless Rodan lime resulting in a fewer number of seeds [82]. However, the fruit set of *Citrus limettoides* was better in open-pollinated flowers than emasculated or hand-pollinated flowers [83]. Further, the flower-fruit ratio and fruit-seed ratio of lemons declined in the absence of pollinators than in open-pollinated conditions [80]. The yield is further enhanced in lemon with the planting of a suitable pollinizer. For instance, the yield of 'Kagzi Kalan' lemon is supposedly enhanced with the plantation of 10% pummelo along with it [84].

Studies have reported that *Citrus limon* L. was self-incompatible and fruit and seed sets were greater in cross-pollination than self-pollination [85]. Further, it has been observed that pollination by bees is not required for the production of fruits [86]. Further, lemons were reported to be self-compatible that produce fruit by self-pollination without the influence of pollinators [22]. Emasculated unpollinated flowers of *C. limon* produced seedless fruits, indicating its parthenocarpic ability [87]. Although cross-pollination helped *C. limon* produce fruits with germinable seeds, it was self-pollination that produced seedless fruits [87]. However, studies also demonstrated that lemon trees caged without bees produced 42.5% less than open-pollinated trees while trees caged with bees produced 10% less fruit indicating that bees contribute to geitonogamous self-pollination [88].

3.6 Breeding system of citrons

Citron is reported to be receptive to the pollen of the same species [89]. Citrons are presumably self-compatible and hence citrons as maternal parents are not common [89, 90]. Citrons have the lowest heterozygosity which results in their high rate of selfing [91]. It has been observed that the male and female organs of Corsican citrons mature simultaneously before the opening of the petals which indicates its self-fertilization mechanism [92]. As a result, citron propagation by seed in Israeli home gardens seldom results in hybrid seedlings [93]. Further, it has also been reported that no pollination problems have been observed on citron [22].

4. Conclusion and future direction

4.1 Status of pollination study and its relevance in *Citrus*

There is enough evidence to show that the global pollinator population is declining which is harming the agricultural productivity of the crop. *Citrus* is an important cash crop and has a varying degree of pollination requirements. Although, several studies have been conducted to ascertain the pollination requirement of *Citrus,* we found uncertainty in the importance of pollination in *Citrus* as far as market demands are concerned. There are several varieties of *Citrus* [23] that are poorly known from the pollinators and pollination requirement [76]. While McGregor [22] stated that few studies have been conducted in other groups of *Citrus*, we are in a much better position now than we have been in the past to understand the pollination requirements of the plants. Nonetheless, considering that new cultivars are developed every time, the study will always be in its infancy requiring constant monitoring to understand their breeding system. However, we suggest (i) recognizing the market need for *Citrus* varieties of superior eating and processing quality (ii) determining the relevance of pollinators based on prior demand of the market. For instance, consumers' favor seedless fruits because of their appealing appearance and ease of consumption, while producers prefer them due to their ease of processing. Given its relevance, efforts are being done to develop high-quality seedless fruits [58].

4.2 The way forward

In this review, we have explored commercially important *Citrus* species and tried to ascertain their adaptation to pollination requirements. We have also ascertained the breeding system of some of the commercially important *Citrus* varieties around the globe. However, growing evidence suggests that the pollination requirements of *Citrus* may be arguable, the biggest question that arises is the importance of pollination in the *Citrus* industry.

Further, we found a disparity in studies conducted between *Citrus* varieties probably because of the commercial popularity of some varieties over others. We still need studies to be conducted to document important pollinators and understand the breeding system of more varieties. The ever-increasing development of new cultivars poses challenges in understanding the breeding process of such cultivars and a constant effort to study such cultivars is the major challenge that needs to be addressed. It is also evident that there are substantial losses of pollinators in many regions of the globe, therefore segregating *Citrus* varieties and cultivars based on its breeding adaptation, pollination requirements, and most importantly based on the demand of the market would greatly determine the effect of the loss of pollinators in the productivity of *Citrus*. However, if the lack of pollinators does affect productivity and disrupts its market supply, targeting on development of seedless varieties can be a good alternative. Also, genetically self-incompatible cultivars that are dependent on pollinators need to be consistently investigated and their pollinators identified. Identifying suitable pollinizers can be another way in enhancing the productivity of *Citrus*. However, such a process also requires pollinators, and therefore, given the importance of pollination, efforts should be made in mitigating pollinator loss to sustainable pollination service. This can be achieved by acknowledging their intrinsic value, considering the core importance of pollinators for the productivity of *Citrus*, focusing on agri-environment schemes, protected area networks, and managing alternative pollinators.

Acknowledgements

The authors are grateful to the Department of Botany and the Central library of Sikkim University for providing access to the journals. The review was prepared without any specific grant from funding agencies in the public, commercial, or not-for-profit sectors.

Conflict of interest

The authors declare no conflict of interest.

Author details

Subhankar Gurung and Arun Chettri*
Taxonomy and Biodiversity Lab, Department of Botany, Sikkim University, Gangtok, Sikkim, India

*Address all correspondence to: achettri01@cus.ac.in

References

[1] Potts SG, Imperatriz-Fonseca V, Ngo HT, Aizen MA, Biesmeijer JC, Breeze TD, et al. Safeguarding pollinators and their values to human well-being. Nature. 2016;**540**:220-229. DOI: 10.1038/nature20588

[2] Das A, Sau S, Pandit MK, Saha K. A review on: Importance of pollinators in fruit and vegetable production and their collateral jeopardy from agro-chemicals. Journal of Entomology and Zoology Studies. 2018;**6**(4):1586-1591

[3] Klein AM, Vaissiere BE, Cane JH, Steffan-Dewenter I, Cunningham SA, Kremen C, et al. Importance of pollinators in changing landscapes for world crops. Proceedings of the Royal Society of London B: Biological Sciences. 2007;**274**:303-313

[4] Sundriyal M, Sundriyal RC. Wild edible plants of the Sikkim Himalaya: Nutritive values of selected species. Economic Botany. 2004;**58**:286-299. DOI: 10.1663/0013-0001(2004)058[0286,WEPOTS]2.0.CO;2

[5] IPBES. The Assessment Report of the Intergovernmental Science-Policy Platform on Biodiversity and Ecosystem Services on Pollinators, Pollination and Food Production. Bonn, Germany: Secretariat of the Intergovernmental Science-Policy Platform on Biodiversity and Ecosystem Services; 2016

[6] Greenleaf SS, Kremen C. Wild bees enhance honey bees' pollination of hybrid sunflower. Proceedings of the National Academy of Sciences of the United States of America. 2006; **103**:13890-13895. DOI: 10.1073/pnas.0600929103

[7] Krebs JR, Wilson JD, Bradbury RB, Siriwardena GM. The second silent spring? Nature. 1999;**400**:611-612

[8] Tscharntke T, Kelin AM, Kruess A, Steffan-Dewenter I, Thies C. Landscape perspective on agricultural intensification and biodiversity-ecosystem service management. Ecology Letters. 2005;**8**:857-874

[9] Biesmeijer JC, Roberts SPM, Reemer M, Ohlemüller Edwards M, Peeters T, Schaffers AP, et al. Parallel declines in pollinators and insect-pollinated plans in Britain and the Netherlands. Science. 2006;**313**:351-354

[10] Rathcke B. Competition and facilitation among plants for pollination. In: Real L, editor. Pollination Biology. London, UK: Academic Press; 1983. pp. 305-329

[11] Rodrigo J, Herrero M. Evaluation of pollination as the cause of erratic fruit set in apricot 'Moniqui'. Journal of Horticultural Science. 1996;**71**:801-805

[12] Partap UMA, Partap TEJ, Yonghua, HE. Pollination failure in apple crop and farmers' management strategies in Hengduan Mountains. China: Acta Horticulturae. 2000;**561**:225-230

[13] Corbet SA. Which bees do plants need? In: Matheson A, Buchmann SL, O'Toole C, Westrich P, Williams JH, editors. The Conservation of Bees. London, UK: Academic Press; 1991. pp. 105-114

[14] Williams IH. The dependence of crop production within the European Union on pollination by honeybees. Agricultural Zoological Review. 1994;**6**:229-257

[15] Kevan PG, Philips T. The economics of pollinator declines: Assessing the consequences. Conservation Ecology. 2001;**5**:8

[16] Ghazoul J. Buzziness as usual? Questioning the global pollination crisis. Trends in Ecology and Evolution. 2005; **20**:367-373

[17] van der JP S, Vaage NS. Pollinators and global food security: The need for holistic global stewardship. Food Ethics. 2016;**1**:75-91

[18] Steffan-Dewenter I, Potts SG, Packer L. Pollinator diversity and crop pollination services are at risk. Trends in Ecology and Evolution. 2005;**20**:651-652

[19] Klein AM, Vaissiere BE, Cane J, Steffan-Dewenter I, Cunningham S, Kremen C, et al. Importance of pollinators in changing landscapes for world crops. Proceedings of the Royal Society of London B: Biological Sciences. 2006;**274**(1608):303-313. DOI: 10.1098/rspb.2006.3721

[20] Casas A, Valiente-Banuet A, Rojas-Martinez A, Dávila P. Reproductive biology and the process of domestication of the columnar cactus *Stenocereus stellatus* in central Mexico. American Journal of Botany. 1999; **86**:534-542

[21] Casas A, Otero-Arnaiz A, Perez-Negron E, Valiente-Banuet A. In situ management and domestication of plants in Mesoamerica. Annals of Botany. 2007;**100**:1101-1115

[22] McGregor SE. Insect pollination of cultivated crop plants. In: Agriculture Handbook 496. USA: USDA-ARS; 1976. p. 395

[23] Webber J, Batchelor LD. The Citrus Industry. 2nd ed. Berkeley and Los Angeles: University of California Press; 1943

[24] Halder S, Ghosh S, Khan R, Khan AA, Perween T, Hasan MA. Role of pollination in fruit crops: A Review. The Pharma Innovation Journal. 2019;**8**(5):695-702

[25] Roubik DW. Pollination of Cultivated Plants in the Tropics. Rome: FAO Agricultural Service Bulletin; 1995

[26] Futch SH, Jackson LK. Pollination of Citrus Hybrids. Florida Cooperative Extension Service IFAS. USA: University of Florida; Fact Sheet No. HS-102. 2003. p. 2

[27] Iwamasa M, Oba Y. Seedlessness due to self-incompatibility in Egami-Buntan, a Japanese pummel cultivar. Agricultural Bulletin of Saga University. 1980;**49**:39-45

[28] Li SJ. Self-incompatibility in 'Matou' wentan (*Citrus grandis* (L.) Osb.). HortScience. 1980;**15**:298-300

[29] Sanford RL, Paaby P, Luvall JC, Phillips E. Climate, geomorphology and aquatic ecosystems. In: McDade LA, Bawa KS, Hespenheide HA, Hartshorn GS, editors. La Selva: Ecology and Natural History of a Neotropical Rain Forest. University of Chicago Press, Chicago, Illinois: USA; 2003. pp. 161-182

[30] Krezdorn AH. Citrus Flowering, Fruit Set and Development. Lake Alfred, FL: Citrus Research and Education Center; 1986

[31] Burger DW. Pollination effects on fruit production of 'Star Ruby' grapefruit (*Citrus paradisi* Macf.). Scientia Horticulturae. 1985;**25**:71-76

[32] Rahman A. Apiculture. In: Kumar AT, Pradhan S, editors. Handbook of Agriculture. Indian Council of Agricultural Research: New Delhi; 2006. pp. 509-529

[33] van Engelsdorp D, Hayes J Jr, Underwood RM, Pettis J. A survey of honeybee colony losses in the U.S., Fall 2007 to Spring 2008. PLos One. 2008;**3**:e4071. DOI: 10.1371/journal.pone.0004071

[34] Potts SG, Roberts SPM, Dean R, Marris G, ABrown M, Jones R, et al. Declines of managed honeybees and beekeepers in Europe? Journal of Apicultural Research. 2010;**49**:15-22

[35] Garibaldi LA et al. Wild pollinators enhance fruit set of crops regardless of honeybee abundance. Science. 2013;**339**:1608-1611

[36] Thompson CE, Biesmeijer JC, Allnutt TR, Pietravalle S, Budge GE. Parasite pressures on feral honeybees (*Apis mellifera* sp.). PLoS One. 2014;**9**:e105164. DOI: 10.1371/journal.pone.0105164

[37] Kraus B, Page REJ. Effect of *Varroa jacobsoni* (Mesostigmata: Varroidae) on feral *Apis mellifera* (Hymenoptera: Apidae) in California. Environment Entomology. 1995;**24**:1473-1480

[38] Hung JKL, Kingston JM, Albrecht M, Holway DA, Kon JR. The worldwide importance of honeybees as pollinators in natural habitats. Proceedings of Royal Society B. 2017;**285**:20172140. DOI: 10.1098/rspb.2017.2140

[39] Thomson D. Competitive interactions between the invasive European honeybee and native bumblebees. Ecology. 2004;**85**:458-470. DOI: 10.1890/02-0626

[40] Thomson DM. Local bumblebee decline linked to recovery of honeybees, drought effects on floral resources. Ecology Letters. 2016;**19**:1247-1255. DOI: 10.1111/ele.12659

[41] Gross CL, Mackay D. Honeybees reduce fitness in the pioneer shrub *Melastoma affine* (Melastomaceae). Biological Conservation. 1998;**86**:169-178. DOI: 10.1016/SOOO6-3207(98)00010-X

[42] Cholis MN, Atmowidi TA, Kahono S. The diversity and abundance of visitor insects on pummel (*Citrus maxima* (burm) Merr) cv. Nambangan. Journal of Entomology and Zoology Studies. 2020;**8**(4):344-351

[43] Free JB. Insect Pollination of Crops. New York: Academic Press; 1970

[44] Zhang S, Liang M, Wang N, Xu Q, Xiuxin D, Chai L. Reproduction in woody perennial Citrus: An update on nucellar embryony and self-incompatibility. Plant Reproduction. 2018;**31**:43-57

[45] Hearn GH, Watkins WG, Bishop RK. Orange nectar and pollen in relation to bee activity. Journal of Economic Entomology. 1942;**35**:321-323

[46] Reece PC, Hearn CJ, Gardner FE. Nova tangelo—an early maturing tangerine hybrid. Proceedings of the Florida State Horticultural Society. 1964;**77**:109

[47] Cameron JW, Frost HB. Genetics, breeding and nucellar embryony. In: Reuther W, Webber HJ, Batchelor LD, editors. The Citrus Industry. Revised ed. Vol. II. Berkely: Div. Agric. Sci., University of California; 1968, Chapter 5

[48] Atawia AR, Abd EL-Latif FM, EL-Badawy HE, Abo-Aziz AB, Abou Rayya MSM, Baiea MHM, et al. Effect of various pollination treatments on yield characteristics and fruit quality of shaddock fruits. International Journal of Scientific & Engineering Research. 2016;**7**(5):1-9

[49] Lacarelle A, Miedzyrzecki C. New contributions on the study of clementines, Morocco. In: Experimentation Fruitiere et Maraichere, Rabat, Morocco. Edition Terre Marocaine ed. France; 1937. p. 22

[50] Mesejo C, Martinez-Fuentes A, Reig C, Agusti M. The effective pollination period in 'Clemenules' mandarin, 'Owari' *Satsuma mandarin* and 'Valenia' sweet orange. Plant Science. 2007;**173**(2):223-230

[51] Hearn CJ, Reece PC. Pollination needs of Page, Lee, Nova and Robinson Citrus hybrids. Citrus Indus. 1967;**48**(4):19-23

[52] Gambetta G, Gravina A, Fasiolo C, Fornero C, Galiger S, Inzaurralde C, et al. Self-incompatibility, parthenocarpy and reduction of seed presence in 'Afourer' mandarin. Scientia Horticulturae. 2013;**164**:183-188

[53] Reece PC, Register PD. Influence of pollinators on fruit set of Robinson and Osceola tangerine hybrids. Florida State Horticultural Society Proceedings. 1961;**74**:64-106

[54] Miwa T. On pollination, fertilization phenomena and problems connected with fruit drop in the hyuganatsu mandarin. Bulletin of the Miyazaki University. Natural Science. 1951;**2**:2-67

[55] Zhgenti SK. Pollination of the Japanese persimmon and mandarin orange. In: Krischunas IV, Gubin AF, editors. Pollination of Agricultural Plants. Russia: Moskva, Gos. Izd-vo SelKhoz Lit-ry; 1956. pp. 193-199

[56] Zavrashvili RM. Importance of nectar in flowers of mandarin orange Unshiu. Soobscenija Akademii Nauk Gruzinskoj. 1967;**45**:1, 205-212

[57] Zavrashvili RM. Influence of bees on the yield of citrus trees on the commercial plantations of Georgia. In: 21st International Apiculture Congress Proceedings College Park, Md. USA. 1967. pp. 450-451

[58] Vardi A, Levin I, Carmin N. Induction of seedlessness in citrus: From classical techniques to emerging biotechnological approaches. Journal of the American Society for Horticultural Sciences. 2008;**133**:117-126

[59] Hearn CJ, Reece PC, Fenton R. Self-incompatibility and the effects of different pollen sources upon fruit characteristics of four citrus hybrids. In: 1st International Symposium Citrus Symposium Proceedings. USA: University of California; Vol. 1. 1969. pp. 183-187

[60] Mustard MJ. Progress Report on the Unfruitfulness of the Minneola Tangelo; Ceiba. 1961;**9**(1):45-48

[61] Ngo BX, Wakana A, Park SM, Nada Y, Fukudome I. Pollen tube behaviours in self-incompatible and self-compatible *Citrus* cultivars. Journal of the Faculty of Agriculture, Kyushu University. 2001;**45**:443-457

[62] Yamamoto M, Kubo T, Tominaga S. Self- and cross-incompatibility of various *Citrus* accessions. Journal of the Japanese Society for Horticultural Science. 2006;**75**(5):372-378

[63] Malerbo-Souza DT, Nogueira-Couto RH, Couto LA. Pollination in orange sweet crop (*Citrus sinensis* L. Osbeck, var. Pera-Rio). Brazilian Journal of Veterinary Research and Animal Science. 2003;**40**(4):237-242. DOI: 10.1590/S1413-95962003 00040001

[64] Malerbo-Souza DT, Nogueira-Couto RH, Couto LA. Honeybee attractants and pollination in sweet orange, *Citrus sinensis* (L.) Osbeck, var. Pera-Rio. Journal of Venomous Animals and Toxins including Tropical Diseases. 2004;**10**(2):144-153

[65] Francke R, Jorge A, Mathieu JM. Effects of insect pollinators on the production of valencia oranges: The honey bee and its effects on citrus production. Agronomia (Monterrey). 1969;**122**:7

[66] Sanford MT. RF-AA092. Pollination of Citrus by Honeybees. Florida Cooperative Extension Service, Institute of Food and Agricultural Sciences. USA: University of Florida; 1992

[67] Koltunow AM, Soltys K, Nito N, McClure S. Anther, ovule, seed and nucellar embryo development of *Citrus sinensis* cv. Valencia. Canadian Journal of Botany. 1995;**73**:1567-1582

[68] Krezdorn AH. Fruit setting problems in Citrus. American Journal of Horticultural Science Carribean Region Proceedings. 1965;**9**(13):85-92

[69] Krezdorn AH. Pollination requirements of Citrus. In: The Indispensable Pollinators. Vol. 127. USA: Arkansas Agriculture Extension Service Misc. Pub.; 1970. pp. 211-218

[70] Atkins EL. Honeybees and agriculture. California Citrograph. 1963;**49**(2):81-82

[71] Wafa AK, Ibrahim SH. Effect of the honeybee as a pollinating agent on the yield of orange. Elfelaha. Egypt: Cairo University; 1960:AA-448/63

[72] Davenport TL. Citrus flowering. Horticultural Reviews. 1986;**12**:349-408

[73] Busch JW, Schoen DJ. The evolution of self-incompatibility when mates are limiting. Trend in Plant Science. 2008;**13**(3):128-136

[74] Aala FT. Effects of Hand pollination on the production of Siamese pummel. Phillipines Journal of Agriculture. 1953;**18**(1-4):101-113

[75] Hoang TT, Nguyen TH, Le HT, Ngo BX, Wakana A. Self-incompatibility in pummel [*Citrus grandis* (L.) Osbeeck] with focus on Vietnamese cultivars with or without parthenocarpy. Journal of the Faculty of Agriculture, Kyushu University. 2014;**59**:65-70

[76] Chacoff NP, Aizen MA. Pollination requirements of pigmented grapefruit (*Citrus paradisi* Macf.) from Northwestern Argentina. Crop Science. 2007;**47**:1-9

[77] Krezdorn AH. Pollination requirements of Citrus. Citrus Indus. 1972;**53**:5-7, 28

[78] Wright N. Pollination and the seediness of marsh grapefruit. Proceedings of the Agricultural Society of Trinidad and Tobago. 1937;**37**:51-60

[79] Crane E, Walker P. Pollination Directory for World Crops. Bucks, UK: International Bee Research Association; 1984

[80] Karmakar P. Pollination biology of *Citrus aurantiifolia* (Christm.) Swingle: A medicinally important fruit plant. International Journal of Innovative Research and Development. 2013; **2**(7):138-142

[81] Soost RK, Cameron JW. 'Oroblanco' a tripliod pummel-grapefruit hybrid. HortScience. 1980;**15**:667-669

[82] Abrahimy Y, Haji VS. Comparison of Minab Lime, Rodan Lime and Persian Lime Cultivars on Different Citrus Rootsocks in Minab, Hormozgan, Iran. Bandar Abas (Iran): Iran Citrus Research Institute—ICRI; 2004. Available from: http://www.agrisis.org

[83] Motial VS. Fruit-set studies in sweet lime. Indian Academy of Science Proceedings Section B. 1964; **60**(6):371-379

[84] Haldhar SM, Kumar R, Samadia DK, Singh B, Singh H. Role of insect pollinators and pollinizers in arid and semi-arid horticultural crops. Journal of Agriculture and Ecology. 2018;**5**:1-25

[85] Hossain MI, Rabbani MG. Study on the cross compatibility of some lemon genotypes (*Citrus limon* L.). Bangladesh Journal of Agriculture Research. 2011;**36**(2):241-246

[86] Webber HJ. Influence of pollination on set of fruit in citrus. California Citrog. 1930;**15**(7):304:322-323

[87] Zhang SW, Huang GX, Ding F, He XH, Pan JC. Mechanism of seedlessness in a new lemon cultivar

'Xiangshui' [*Citrus limon* (L.) Burm, F.]. Sex Plant Reproduction. 2012; **25**:337-345

[88] Zavrashvili RM. Bees and the citrus crop. Pchelovodstvo. 1964;**84**(8):19

[89] Moore GA. Oranges and lemons: Clues to the taxonomy of *Citrus* from molecular markers. Trends in Genetics. 2001;**50**:670-672

[90] Scora RW. On the history an origin of *Citrus*. Bulletin of the Torey Botanical Club. 1975;**102**:369-375

[91] Barkley NA, Roose ML, Krueger RR, Federici CT. Assessing genetic diversity and population structure in a citrus germplasm collection utilizing simple sequence repeat markers (SSRs). Theoretical and Applied Genetics. 2006;**112**:1519-1531

[92] Luro FN, Venturini G, Costantino J, Paolini J, Ollitrault P, Costa J. Genetic and chemical diversity of citron (*Citrus medica* L.) based on nuclear and cytoplasmic markers and leaf essential oil composition. Phytochemistry. 2012;**77**:186-196

[93] Goldschmidt EE. Citron propagation: Seeds or cuttings? Alon Hanoeta. 2009;**63**:44-46. (Hebrew)

Chapter 5

Invasive Exotic Plant-Pollinator Interactions

B.C. Anu and Ramanuj Vishwakarma

Abstract

Pollination is an imperative biological process, and the exotic plant species have a significant effect on the interaction of plant with pollinators. The exotic plant communities have the ability to cause both direct and indirect impacts on pollinators. The impact of non-native exotic plants on native pollinators can occur at a varying range of scales: starting from the flower visitors who visit flowers individually, to populations and community-level interactions (insect-flower interaction networks). As it is impractical to study every invasive plant in every ecological context, understanding appropriate individual-level trait predicting direct interactions between invasive exotic plants and native pollinators is needed.

Keywords: community-level interactions, ecological context, exotic plant, flower visitors, invasive, non-native

1. Introduction

Pollination that is dependent on an insect is essential in many plants for seed and fruit production, and flowers offer the nectar and pollen that are required for the development of many insects. A lot of attention is gained from ecologists on plant-pollinator interactions, which been perceived as robust architectural property networks [1]. The networks between plant and pollinators are highly enclosed, and they consist of generalist species that interact with both generalist and specialist species; there is a connection between this generalist and specialist species structure [2]. It has been claimed that this kind of structure is often environmentally stable along with low sensitivity to sampling effort [3]. The proof of stability in the plant-pollinator network is fright in front of global change as it is mostly based on the species removal models [4] and destruction of habitats [5]. There is a frail and unsymmetrical interaction between plants and pollinators, that is, it is the pollinator species on which plant species depend and that pollinator species weakly depend on reciprocal plants and *vice versa* [6].

Biological invasions can definitely serve as natural community experiments [7]. We can test whether there are disrupting effects of a new species to a community upon insertion on the network of plant and pollinator by comparing communities that are invaded naturally and communities that are uninvaded or by investigating the invasion activities. It has been suggested that there is an efficient use of pollinators that are native, upon the establishment of many invader plants that are generalist and entomophilous [8]. Moreover, the main harmful impact of non-native plant species on native plant pollinators can be demonstrated through plant-pollinator interaction analysis [9].

There may be a negative effect of invasive alien plant species on plant-pollinator networks such as shading because globally the shrubs compose 23% of invasive plant species [10]. There may be an adverse effect of exotic invasive plants on the ecosystems that they invade, leading to a loss in biodiversity and changes in the functioning of the ecosystem [11]. But, so far this theory has not been supported universally by the studies; the results are dependent on the surroundings and vary on the basis of the traits of the invaded community and the invaders [12].

Study has shown the negative impact on the reproduction of native species by the invasive species, especially when the latter is more abundant [13]. The direct effect of non-native invader plants on native pollinators and their role in the invasion of plants has been focused on in some studies. These direct interactions may affect the fitness of both pollinators and the invasive plant species. Keeping in view the above facts, we have tried to wrap information on the impacts of exotic plant species on native pollinators.

2. Role of pollinators in promoting non-native plant invasion

We know very little about the pollination activities of invasive pollinators in their native range that are present worldwide; whether the pollinator species or the similar groups same as pollinators exist in new areas; in the new areas, flowers are visited by which pollinator species; whether there are limited pollinators; how much the pollinator service quality is important; or whether through other means of reproduction they are successful. There has been a study of a few species in the ecology of the pollination in both exotic invasive and native ranges; for example, in Spain, *Hedysarum coronarium* is native on the mainland and to Balearic islands where it has been introduced [14]; *Rhododendron ponticum* is native to Spain, and it is introduced in Ireland where it is highly invasive [15]; and *Nicotiana glauca* is native to South America, and it is globally introduced [16]. In all the cases of the invasive non-native plant species, the native pollinators that are resident and belong to the same functional group as the non-native plants in their native ranges do the successful pollination. The pollination of *H. coronarium* in native, as well as nonnative areas, is done by the same species (Hymenoptera and Coleoptera), particularly, the honeybee *Apis mellifera* [14]; in Spain, large bees *Xylocopa violacea* and *Bombus* spp. pollinate *Rhododendron ponticum* and in Ireland, *R. ponticum* is pollinated by *Bombus* spp. [15]; and hummingbirds pollinate *N. glauca* in the native range; and in the invasive range, specialized sunbirds pollinate *N. glauca* [16].

Nonetheless, knowledge of the ecology of pollination in most species is far more complete in non-native and native regions, as well as for those receiving a great amount of scientific and public attention. For example, in the northern temperate regions, pollination in a notorious invader *Impatiens glandulifera* is done by a variety of bee species in its non-native invasive range [17], but in native India, Pakistan, and Nepal, little is known about its pollinators except for food it is visited by *Bombus* spp. [4]. The role of pollinators in promoting invasion in non-native ranges is poorly studied. Theoretically, the establishment and spread of invasive plants are dependent on the replacement of lost mutualists (including pollinators) from a plant's native range with new mutualist partners in the non-native range [8]. It follows that the self-compatible invasive plant species that can reproduce through asexual propagation and self-pollination plants would be better invaders, as they are less likely to experience a limitation in pollination (i.e., reduction in quantity and quality of pollination service in the new habitat) [18].

Although, through the studies on the invasive plant, characters have revealed that many invasive species are pollinated by biotic agents [19]; but in the case of certain species, self-pollination in high levels can enhance successful reproduction as well as invasion, in a similar way visit by generalist pollinators can also boost the success in reproduction as well as invasion [20]. The pollination service quality is not guaranteed even if appropriate pollinators are present. Pollinators may lack the required phenology or behavior to provide high-quality pollination, resulting in heterospecific or low-quality pollen transfer. In order to become successful invaders, non-native plants, which are pollinated by insects, have several options available. Their pollinators can be introduced alongside them, either simultaneously or later. For example, *Ficus* spp. became invasive in both Florida and New Zealand after the introduction of fig wasps that are host-specific, and viable seed production began [8]. In other cases, pollination may get easier by introducing generalist pollinators. There can be reasons for it, such as either due to the nonexistence of suitable species of pollinator that are native or may be due to the presence of a preexisting adaptation promoting the interaction between plant and pollinator. For example, the introduction of European honeybee (*A. mellifera*) and bumblebee (*Bombus* spp.) to New Zealand introduced European plants as they are frequent pollinators of European plants, for example, *Trifolium pratense* and *Echium vulgare* [21]. Similarly, in North America, introduced *A. mellifera* is a major pollinator of invasive European/Eurasian species *Lythrum salicaria*, *Cirsium vulgare*, and *Rosa multiflora* [22], and in India, *A. mellifera* was primarily introduced as major honey-producing species that becomes later a major pollinator of a number of oilseed crops *viz.*, *Brassica* spp. [23] and *Helianthus annuus* [24], a second important pollinator of *Mangifera indica* [25] and also observed as least visitors of *Capsicum annuum* in the absence of other important floras [26]. Given the broad range of agricultural areas in which *A. mellifera* and *Bombus* spp. are introduced for crop pollination worldwide, they have the ability to promote invasion by a wide range of plant species with which they have coevolved. Pollinators already present in the new habitat can form novel interactions with non-native plants, co-opting them into pollination roles [19]. In many cases, pollinators are frequently native generalists of the same functional group or species as same as the pollinators in the native range of plants, such as *H. coronarium*, *Rhododendron ponticum*, and *N. glauca*.

In other instances, the non-native pollinators that are generalists have coevolved with the close relatives of the non-native plants rather than the non-native plant itself. Again, some good examples of this are the generalist bees that are introduced for crop pollination, such as *Apis* and *Bombus* spp. In Australia, introduced bees (*A. mellifera* and *Bombus terrestris*) pollinate *Lupinus aboreus* [27], and it does not occur in the native range of the species but to the species that are functionally similar to the native pollinators of *L. arboreus* and in Europe, they pollinate native *Lupinus* species. There are also cases where pollination by resident species is done for the relatively specialized pollination system of the plants; for example, in South Africa, *Lilium formosanum*, a non-native invasive species, are pollinated by a widespread hawk moth pollinator [28].

Some of the plant species that are pollinated biotically have become invasive without the need to reproduce sexually and have spread by vegetative means across their invasive range. For example, in parts of its European range, the invasive *Fallopia japonica* entirely relies on the clonal spread, while it reproduces sexually in North America [29]. This phenomenon has been caused by a lack of sexual partners (male individuals that are fertile) rather than a lack of pollinators in this case. *Oxalis pescaprae* is another invasive species that spreads *via* clonal bulb development in parts of its Mediterranean Basin range (**Figure 1**) [30].

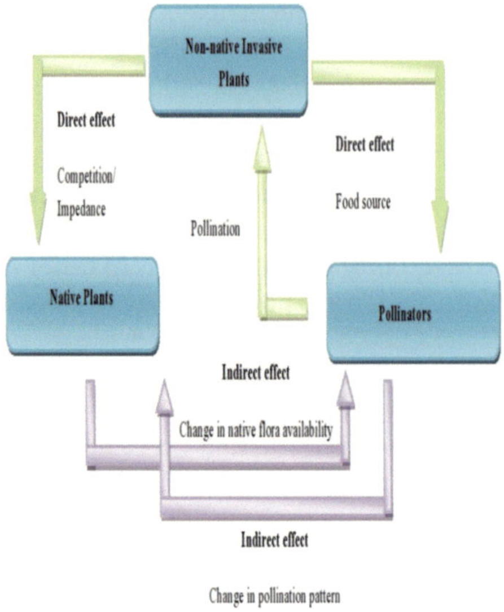

Figure 1.
Interaction of non-native invasive plants with native plants and pollinators [15].

3. Impact of invasive plants on flower visitors

Although interactions with native pollinators may benefit invasive plants, the effect of invasive plants on native flower visitors is thought to be the opposite [31]. This may be because non-native invasive plants take up space that would otherwise be used by native plants, which are thought to be more likely to provide a suitable resource. Invasive plants can also alter the behavior of pollinators that are highly competitive, resulting in indirect effects that can be added on some taxa of pollinators or flower visitors. Despite the fact that quantifying the effects of invasive plants on native arthropods is more complicated than quantifying the effects of native plants [32], due to their mobility, size, and sometimes taxonomic limitations and on the effects of invasive non-native plants on non-pollinating taxa on which several studies have been conducted [31].

There is a lack of knowledge about how arthropod habitat and dietary requirements are affected by invasive plants in general, not just for insects that visit flowers [31]. Most studies on the impacts of an invasive plant on native flower visitors have focused on the species diversity, abundance, and community composition of taxa in non-native plant invaded sites; very little has been explored on the impacts at the individual level or population level. Conflicting conclusions can be seen from the studies that looked into the impact of invasive plants on pollinator abundance. For example, in the case of non-native plants, some authors have found that the invasive plants increase the abundance of certain species, such as generalist butterflies [33] and bees [34], while others have found that invasive plants reduce the abundance of butterflies [35], bees [36], and entire populations of pollinator [37].

Some research by Moron et al. (2009) and Bartomeus et al. (2010), on the other hand, has found no effect on pollinator abundance. Despite the fact that the impact of invasive plants on pollinator species richness and diversity seems to be more consistently negative [37], several studies have found little improvement in species richness after invasion [38]. Invasive plants have been found to influence not only the number of species but also the composition of populations in terms of the taxa present and the presence of a relative abundance of individuals in each taxon, according to several studies. *Vincetoxicum rossicum*, an invasive vine, in general, supported a low number of arthropods, with some feeding guilds completely absent and with the presence of very few pollinators [32]. Given the potential of invasive plants to affect the relative abundance of different taxa, and thus the composition and diversity of pollinator species, several recent studies have attempted to characterize these changes by analyzing whether invasive plants cause functional changes in ecosystems.

Examining the flower-visitor interaction network structure in invaded environments has become the most common method of characterizing changes in the interactions between plant and pollinator. Though non-native plants have become well integrated into interaction networks [39, 40], network structure metrics are sometimes unaffected [41]. Invasive species, on the other hand, can dramatically alter the structure of interaction networks, either by changing the evenness of the interaction and density of linkage (the number of weighted links per species) [42] or within the network by forming larger, more linked modules [43]. The extent of effects on the network structure is likely to be influenced by the relative abundance of invasive species in the network [44]. Given the extent of effects on network structure, the ambiguity surrounding the functional consequences of network structural properties, as well as the ability to rewire networks in the presence of invaders [45]. For native animals, invasive plants may provide an alternative food source, but only if native animals can get access (i.e., depending on the visitors and flowers of the invasive species trait complementarity) [22] and only if the food is valuable that too nutritionally and does not harm the health and fitness of those who eat it.

4. Methods to quantify impacts on native pollinators

The effects of invasive plants on native flower visitors have been studied using a variety of methods. To simulate invasion and examine ecological consequences, cut branches or potted plants have been used in some cases [18]. The majority of studies have been "natural experiments, comparing the effects of invasive plants with varying degrees of separation between areas where they are established and growing naturally to areas where they are not with varying degrees of separation between areas [46, 47]. Both methods have advantages and disadvantages: In the first, the effects of a sudden introduction of a new invasive flowering plant, as well as flower visitor response, can be tracked. Effects on the visitors of the flower, on the other hand, are more likely to be behavioral responses to the invader than responses at the community level.

There have been many approaches to design "natural experiments" in the second case. Within a single location, the preference of flower visitors can be determined by comparing visits to invasive and native flowers that are both present on the same site. Within a site, visitation to both non-native invasive and native flowers could be recorded, and this could be done by comparing the visitation to native flowers by the pollinators in the absence of the invader. The most common use of this method is to investigate the effects on interaction networks through invasion [46]. The majority of studies have compared the abundance/richness of taxa at invaded vs. control sites, or invasive vs. noninvasive (control) plants [7]. Another strategy is to exclude the invader from a site and investigate the impact on the flower visitor population associated with the remaining native plants in that site, or compare it to populations in invaded or uninvaded sites [40].

An alternative method of determining whether flower visitors are affected by an invasive plant because of its invasive status is to compare the role of the plant in invasive ecosystems versus native ecosystems. This method has not been used often, but it may be useful in assessing impacts because an attractive plant species that dominates interaction networks in its native habitat may have similar effects when it is not native.

Thus, investigating invasive species through biogeographical studies that look at both their native and introduced ranges may provide valuable information about invasion ecology. However, since many of the effects on flower visitors of invasive plants may be due to the abundance of flowers they grow on and thus are highly rewarding, a different approach would be to compare the effects of invasive plants with the effects of highly rewarding native plants. Then, rather than just the effects of a highly rewarding species in a system, we may be able to separate the effects of non-native invasive plants on flower visitors as a whole.

5. Conclusions

Interaction is often formed between invasive plants and native flower visitors, and their invasion is often reliant on these interactions. Despite the widespread belief that invasive plants have a detrimental effect on native flower tourists, there is no evidence to back this up [48]. In view of the challenges of studying so many flower-visiting taxa, it is not surprising that the direct effects on the diet of individual flower visitors, their health, and fitness by the invasive plants are poorly understood. Given the complexity of performing experiments to assess the effects and the wide variety of responses that can be measured, we are only likely to be able to evaluate impacts for a small percentage of the potentially affected species (most likely insect species that are commercially accessible and/or can be manipulated in a laboratory setting). Moreover, since the effects of invasive plants are likely to be specific to plant species and ecological background-specific, our knowledge is likely to be restricted to globally distributed, problematic plant species. However, better predictions of impacts can be made by designing suitable studies [49] and integrating more knowledge of plant and insect species traits (including plant the breeding system, pollination syndrome, nectar chemistry, insect body size, and diet breadth). As a result, more research into invasive plants and the ecology of flower visitors is needed before broad conclusions regarding direct impacts can be drawn.

Acknowledgements

We take this opportunity to thank every person who supported us in the preparation of this manuscript. We shall be forever grateful to our institution, that is, Bihar Agricultural University, Sabour, Bhagalpur, Bihar, India, for providing us a scientific forum for our holistic development.

Conflict of interest

The authors declare no conflict of interest.

Author details

B.C. Anu and Ramanuj Vishwakarma*
Department of Entomology, Bihar Agricultural University, Bhagalpur, India

*Address all correspondence to: entoramanuj@gmail.com

References

[1] Bascompte J, Jordano P, Olesen J. Asymmetric coevolutionary networks facilitate biodiversity maintenance. Science. 2006;**312**:431-433

[2] Blüthgen N, Menzel F, Hovestadt T, Fiala B, Blüthgen N. Specialization, constraints, and conflicting interests in mutualistic networks. Current Biology. 2007;**17**:341-346. DOI: 10.1016/j.cub.2006.12.039

[3] Nielsen A, Bascompte J. Ecological networks, nestedness and sampling effort. Journal of Ecology. 2007;**95**:1134-1141. DOI: 10.1111/j.1365-2745.2007.01271.x

[4] Saini M, Raina R, Khan Z. Taxonomy and pollination ecology of *Bombus rufofasciatus* (Hymenoptera: Apidae) from the Indian Himalaya. Polish Journal of Entomology. 2013;**81**:347-363

[5] Fortuna MA, Bascompte J. Habitat loss and the structure of plant-animal mutualistic networks. Ecology Letters. 2006;**9**:281-286. DOI: 10.1111/j.1461-0248.2005.00868.x

[6] Petanidou T, Kallimanis AS, Tzanopoulos J, Sgardelis SP, Pantis JD. Long-term observation of a pollination network: Fluctuation in species and interactions, relative invariance of network structure, and implications for estimates of specialization. Ecology Letters. 2008;**11**:564-575. DOI: 10.1111/j.1461-0248.2008.01170.x

[7] Sargent RD, Ackerly DD. Plant-pollinator interactions and the assembly of plant communities. Trends in Ecology & Evolution. 2008;**23**:123-130

[8] Richardson DM, Allsopp N, D'Antonio CM, Milton SJ, Rejmanek M. Plant invasions—The role of mutualisms. Biological Reviews of the Cambridge Philosophical Society. 2000;**75**:65-93. DOI: 10.1017/S0006323199005435

[9] Morales CL, Traveset A. A meta-analysis of impacts of alien vs. native plants on pollinator visitation and reproductive success of co-flowering native plants. Ecology Letters. 2009;**12**:716-728. DOI: 10.1111/j.1461-0248.2009.01319.x

[10] Martin PH, Canham CD, Marks PL. Why forests appear resistant to exotic plant invasions: Intentional introductions, stand dynamics, and the role of shade tolerance. Frontiers in Ecology and the Environment. 2009;**7**:142-149

[11] Mack RN, Simberloff D, Lonsdale WM, Evans J, Clout M, Bazzaz FA. Biotic invasions: Causes, epidemiology, global consequences, and control. Ecological Applications. 2000;**10**:687-710

[12] Pysek P, Jarosık V, Hulme PE, Pergl J, Hejda M, Schaffner U. A global assessment of invasive plant impacts on resident species, communities and ecosystems: The interaction of impact measures, invading species' traits and environment. Global Change Biology. 2012;**18**:1725-1737

[13] Dietzsch AC, Stanley DA, Stout JC. Relative abundance of an invasive alien plant affects native pollination processes. Oecologia. 2011;**167**:469-479

[14] Montero-Castano A, Vila M, Ortiz-Sanchez FJ. Pollination ecology of a plant in its native and introduced areas. Acta Oecologica. 2014;**56**:1-9

[15] Stout JC, Parnell JAN, Arroyo J, Crowe TP. Pollination ecology and seed production of *Rhododendron ponticum* in native and exotic habitats. Biodiversity and Conservation. 2006;**15**:755-777

[16] Ollerton J, Watts S, Connerty S, Lock J, Parker L, Wilson I. Pollination ecology of the invasive tree tobacco *Nicotiana glauca*: Comparisons across native and non-native ranges. Journal of Pollination Ecology. 2012;**9**:85-95

[17] Nienhuis CM, Stout JC. Effectiveness of native bumblebees as pollinators of the alien invasive plant *Impatiens glandulifera* (Balsiminaceae) in Ireland. Journal of Pollination Ecology. 2009;**1**:1-11

[18] Goodell K, McKinney AM, Lin CH. Pollen limitation and local habitat-dependent pollinator interactions in the invasive shrub *Lonicera maackii*. International Journal of Plant Sciences. 2010;**171**:63-72

[19] Pysek P, Jarosık V, Chytry M, Danihelka J, Kuhn I, Pergl J. Successful invaders co-opt pollinators of native flora and accumulate insect pollinators with increasing residence time. Ecological Monographs. 2011;**81**:277-293

[20] Powell K, Krakos K, Knight T. Comparing the reproductive success and pollination biology of an invasive plant to its rare and common native congeners: A case study in the genus Cirsium (Asteraceae). Biological Invasions. 2011;**13**:905-917

[21] Goulson D, Hanley M. Distribution and forage use of exotic bumblebees in South Island, New Zealand. New Zealand Journal of Ecology. 2004; **28**:225-232

[22] Jesse LC, Moloney KA, Obrycki JJ. Insect pollinators of the invasive plant, *Rosa multiflora* (Rosaceae), in Iowa, USA. Weed Biology and Management. 2006;**6**:235-240

[23] Vishwakarma R, Chand P. Foraging activity of insect pollinators and their impact on yield of rapeseed-mustard. BIOINFOLET. 2017;**14**(3):222-227

[24] Vishwakarma R, Ghatak SS. Impact of foraging activity of pollinators including honeybees on seed yield of sunflower. Indian Journal of Entomology. 2014;**76**(2):136-141

[25] Vishwakarma R, Singh R. Foraging behaviour of insect visitors and their effect on yield of mango var. Amrapali. Indian Journal of Entomology. 2017; **79**(1):72-75

[26] Vishwakarma R, Singh R. Insect pollinators and their pollinating efficiency on fruit yield of chilli. Advances in Bioresearch. 2018;**9**(6): 124-129

[27] Stout J, Kells A, Goulson D. Pollination of the invasive exotic shrub *Lupinus arboreus* (Fabaceae) by introduced bees in Tasmania. Biological Conservation. 2002;**106**:42

[28] Rodger JG, van Kleunen M, Johnson SD. Does specialized pollination impede plant invasions? International Journal of Plant Sciences. 2010;**171**:382

[29] Forman J, Kesseli RV. Sexual reproduction in the invasive species *Fallopia japonica* (Polygonaceae). American Journal of Botany. 2003; **90**:586-592

[30] Ferrero V, Barrett SCH, Castro S, Caldeirinha P, Navarro L, Loureiro J. Invasion genetics of the Bermuda buttercup (*Oxalis pes-caprae*): Complex intercontinental patterns of genetic diversity, polyploidy and heterostyly characterize both native and introduced populations. Molecular Ecology. 2015; **24**:2143-2155

[31] Litt AR, Cord EE, Fulbright TE, Schuster GL. Effects of invasive plants on arthropods. Conservation Biology. 2014;**28**:1532-1549

[32] Ernst C, Cappuccino N. The effect of an invasive alien vine, *Vincetoxicum rossicum* (Asclepiadaceae), on arthropod

populations in Ontario old fields. Biological Invasions. 2005;7:417-425

[33] Shapiro AM. The Californian urban butterfly fauna is dependent on alien plants. Diversity and Distributions. 2002;**8**:31-40

[34] Tepedino VJ, Bradley BA, Griswold TL. Might flowers of invasive plants increase native bee carrying capacity? Intimations from Capitol Reef National Park, Utah. Natural Areas Journal. 2008;**28**:44-50

[35] Hanula JL, Horn S. Removing an exotic shrub from riparian forests increases butterfly abundance and diversity. Forest Ecology and Management. 2011;**262**:674-680

[36] Fenesi A, Vagasi CI, Beldean M, Foldesi R, Kolcsar LP, Shapiro JT. *Solidago canadensis* impacts on native plant and pollinator communities in different-aged old fields. Basic and Applied Ecology. 2015;**16**:335-346

[37] Moron D, Lenda M, Skorka P, Szentgyorgyi H, Settele J, Woyciechowski M. Wild pollinator communities are negatively affected by invasion of alien goldenrods in grassland landscapes. Biological Conservation. 2009;**142**:1322-1332

[38] Bartomeus I, Vila M, Steffan-Dewenter I. Combined effects of *Impatiens glandulifera* invasion and landscape structure on native plant pollination. Journal of Ecology. 2010;**98**:440-450

[39] Lopezaraiza-Mikel ME, Hayes RB, Whalley MR, Memmott J. The impact of an alien plant on a native plant-pollinator network: An experimental approach. Ecology Letters. 2007;**10**: 539-550

[40] Traveset A, Olesen JM, Nogales M, Vargas P, Jaramillo P, Antolın E. Bird-flower visitation networks in the Galapagos unveil a widespread interaction release. Nature Communications. 2015;**6**:6376

[41] Vila M, Bartomeus I, Dietzsch AC, Petanidou T, Steffan-Dewenter I, Stout JC. Invasive plant integration into native plant-pollinator networks across Europe. Proceedings of the Royal Society B: Biological Sciences. 2009;**276**:3887-3893

[42] Kaiser-Bunbury CN, Valentin T, Mougal J, Matatiken D, Ghazoul J. The tolerance of island plant-pollinator networks to alien plants. Journal of Ecology. 2011;**99**:202-213

[43] Albrecht M, Padron B, Bartomeus I, Traveset A. Consequences of plant invasions on compartmentalization and species' roles in plant-pollinator networks. Proceedings of the Royal Society B: Biological Sciences. 2014;**281**: 20140773

[44] Stout JC, Casey LM. Relative abundance of an invasive alien plant affects insect-flower interaction networks in Ireland. Acta Oecologica. 2014;**55**:78-85

[45] Campos-Navarrete MJ, Parra-Tabla V, Ramos-Zapata J, Dıaz-Castelazo C, Reyes-Novelo E. Structure of plant-Hymenoptera networks in two coastal shrub sites in Mexico. Arthropod-Plant Interactions. 2013;7:607-617

[46] de Groot M, Kleijn D, Jogan N. Species groups occupying different trophic levels respond differently to the invasion of semi-natural vegetation by *Solidago canadensis*. Biological Conservation. 2007;**136**:612-617

[47] Jakobsson A, Padron B, Traveset A. Competition for pollinators between invasive and native plants: Effects of spatial scale of investigation (Note). Ecoscience. 2009;**16**:138-141

[48] Traveset A, Richardson DM. Biological invasions as disruptors of plant reproductive mutualisms. Trends in Ecology & Evolution. 2006;**21**: 208-216

[49] Kumschick S, Gaertner M, Vila M, Essl F, Jeschke JM, Pysek P. Ecological impacts of alien species: Quantification, scope, caveats, and recommendations. Bioscience. 2015;**65**:55-63

Chapter 6

Reproductive Ecology of Forage Alfalfa (*Medicago sativa* L.): Recent Advances

Hana'a Burezq

Abstract

Plants display an assorted collection of reproductive tactics that eventually play a crucial role in perpetuation of species. Plant reproductive ecology is principally concerned with the adaptive implications of the plant in their vicinity, disparity in qualities allied with pollination, seed dispersal, and seedling establishment. The success in reproduction in most flowering plants depends on ecological interactions with pollinators and seed dispersal agents. Modern tactics in reproductive ecology can integrate proper surveys, advanced pollination studies, interaction between flower and pollinators and clear assessments of population genetic structure, which can provide new opportunities for plant reproductive biology. Alfalfa is an important forage legume and known as "Queen of forages" due to its worldwide adaptability, high yield potential and quality. Alfalfa produces seeds which are primarily used for forage production. It is a gift to livestock industry including dairy, beef, horses, and sheep for grazing, silage, hay etc. Alfalfa is also a medicinal herb with antioxidant, antidiabetic, anti-inflammatory, neuroprotective and cardioprotective properties, utilized for treatment of arthritis, kidney problems. The seeds are exploited in alfalfa sprout industry. The current chapter highlights the reproductive biology of alfalfa from flower development to seed production and its advances.

Keywords: alfalfa, reproductive ecology, forage, seeds

1. Introduction

Reproduction is the critical phase in the life time of any organism, which ensures its perpetuation as a discrete species. The accomplishment of the phenomenon is greatly influenced by two components *viz.,* intrinsic reproductive biology of the organism and the interactive niche in which they survive. The environment in which the plants grow has a major impact as it is decisive in every step of reproduction in plants, due to the fact that the plants are immobile [1]. Therefore, the reproductive success of plant can be constrained by any attributes in the ecosystem where they grow. Alfalfa (*Medicago sativa* L.), commonly called as Lucerne in many other countries is the most important forage crop accounting for the major load of hay produced globally; recognized as the back bone for many forage systems. In Gulf region, alfalfa is widely cultivated and is recognized as the first cultivated forage crop occupying around 30% of the cultivated area [2]. Alfalfa produces nutrient dense foliage with great palatability and digestibility to the cattle. Alfalfa yields better than most forages under less ideal environmental conditions due to its deep root system. Besides, it has long been used medicinally for treatment of arthritis, kidney problems and has antioxidant, antidiabetic, anti-inflammatory, neuroprotective and cardioprotective properties. The factors that impede with the reproductive ecology of alfalfa from reproduction through pollination, seed production and dispersal and the recent advances are highlighted in the current review.

1.1 Plant reproductive biology

Reproductive awareness of any plant is vital for understanding its seed production and dispersal for the continuation of the generation [3]. Plants that do not reproduce vegetatively, depend on seeds to produce offsprings. To be established into a new generation, plants must be pollinated to form fruits, ovules must be fertilized to form viable seeds and the seeds must be dispersed to suitable substrates for germination and growth [4]. Any weak link or break in the chain of events restricts a plant's ability to reproduce and over constant time, contribute to deprivation and impedes the continuation of generation, thereby conservation [5]. Plant reproductive biology may encompass the pollination mechanisms, seed set and seed dispersal between and within populations. Consequently, a comprehensive acquaintance in reproductive knowledge of any plant is mandatory to develop strategies for its conservation [6]. Reproductive biology is a science dealing with the study of the relations between plants and pollinators as well as seed or fruit dispersers. Pollination is an indispensable event in the sexual life of flowering plants, signifying a biotic mutualism between flora and fauna [7]. It is the decisive phase for successive reproduction and suitability of flowering plants for maintenance of seed set and genetic diversity. Pollinators play a key role in reproduction of all flowering plants in the globe, especially agricultural crops depend on pollinators for successful crop production [8]. More than 87% of flowering plants are pollinated by pollinators and insects constitute the major group among pollinators. Plants and pollinators have coevolved to progress mutualistic adaptations, with large ecological changes that dissociate the coincidence in flowering and breeding cycles of pollinators. These coevolutionary relationships are now being severely affected due to global decline in pollinators caused by environmental and anthropogenic disturbances [9]. The human induced interferences such as habitat loss due to fragmentation, deforestation, irrational agricultural practices, extensive use of harmful chemical pesticides, etc. tend to unfasten the advantages of mutual adaptableness of plants and pollinators. The ecological impacts such as pollution and climate change also cause severe threat. Limited pollen or nectar production in plants or ineffective pollinator services due to environmental changes are the chief reasons for the fewness of several important medicinal plants, tree crops etc. [10]. Decline in pollination leads to reduced fruit and seed set, thereby curbing regeneration and limit fitness of plants. Fruit or seed dispersals is vital event for a plant to extend its generation to diverse localities including wild habitats. Thus, the ecoservices by pollinators and dispersal agents errand a crucial part in success of reproduction in a plant.

Observations and experimentation on assorted areas of pollination biology and seed dispersal ecology are a must to comprehend the countless intricacies dominant in plant-pollinator as well as seed dispersal systems [11]. The awareness in this area is a prerequisite for the conservation and management of plant and their associated animal species. Research in reproductive biology of plants will aid to develop new strategies to preserve the genetic potential of the plant species, which are crucial for restoration and reintroduction.

1.2 Plant reproductive ecology

Plant reproductive ecology has arisen as the most adaptable and important fields of plant sciences. It contracts with the reproduction of plants to sustain their generation and the ecological strategies that aid to overcome reproductive catastrophe [12]. For reproductive success of plants owing to its tremendous diversity and ecological niches, substantial amount of disparity is anticipated in the ecological processes [13]. The data base of reproductive biology includes the facts of flowering phenology, floral biology, pollination mechanism, pollen-pistil mechanism, breeding system and natural recruitment.

Lately, the advancement in the particular subject has been stimulated by intensifying interdisciplinarity to address the chief global concerns of warranting food security and biodiversity conservation in the light of several hindrances in plant-pollinator-dispersal agent mutualisms [14]. In this view, the current review describes the reproductive biology of the chief forage plant, alfalfa, from flowering to seed dispersal to enhance the baseline data on its biology and identify deficiencies in its capacity to reproduce and the latest advances to overcome it [15]. The main focus is to synthesize and utilize the database for greater understanding and clarification of the ecological processes governing reproduction.

2. Alfafa: *M. sativa* L.

Alfalfa (*M. sativa* L.) is recognized as the oldest forage grown worldwide, due to its high feeding value and wide adaptability [16]. It is a forage crop with several benefits by strengthening soil structure, contribute nitrogen to soil as well as pest management impacts. Alfalfa is cultivated in more than 80 countries around the globe in an area of 32-million-hectare [17], with a forage yield of 5.75 metric tonnes per hectare per year, hay yield of 8.3 million metric tonnes per year and seed yield of 186–280 kg ha^{-1} annually [18], in addition to the contribution of 83–594 kg N ha^{-1} with the aid of Nitrogen fixing bacteria, *Sinorhizobium meliloti* in the root nodules [19]. Alfalfa is a highly nutritious animal feed and preferred to other forages by the feeding ruminants [20]. The potentiality of alfalfa as a cattle feed arises due to the high content of crude protein with excellent digestibility and its rapid passage through the gastro intestinal track, providing quality proteins for gut microorganisms to resynthesize proteins and vitamins, and stimulate cellulose digestion [21]. The appropriate predicted time of harvest is 10% blooming stage for higher biomass as well as forage yield [22]. As flowering is onset, several physiological functions occur in alfalfa plant. A chief reorganization of photosynthetic possessions from foliage to reproductive parts [23], an increase in lignification of tissues occurs [24], the quality of stem tissues declines and leaf tissues improves [25]. As flowering onsets, a loss in foliage quality up to 45% in relation to feed value is documented. Moreover, a per cent loss in foliage quality leads to 3% loss of cattle daily weight gain [26]. Therefore, a delay in flowering leads to high forage quality for longer period.

2.1 Propagation of alfalfa

Seed is the base for any plant to propagate. Alfalfa is one of the chief forage plants cultivated in dry region through seeds [27]. Seeds are produced in alfalfa for propagation as well as for sprout production for human consumption. Alfalfa can be propagated sexually by seeding alfalfa seeds in a well prepared and well drained seed bed @8–10 kg per acre as pure crop; for mixed strands with grasses, a seed rate of 6–8 kg per acre is required. Asexually alfalfa can be propagated by semi rigid stem cuttings.

2.1.1 Vegetative growing stages of Alfalfa

This interactive diagram demonstrates the different growing stages of alfalfa.

2.1.1.1 Germination and seeding emergence

Seeds should be planted half an inch deeper in the soil in a well-prepared soil, supplied with sufficient nutrients. The alfalfa seeds will absorb water about 24–48 h after planting, provided adequate water is present in the soil. As the seed imbibe water, a radicle root emerges forming the taproot that anchors the seed to the soil surface. The hypocotyl which is the initial seedling develops by pulling the seed coat up through the soil [28]. Germination takes place at optimum temperature range of 65–72°C (**Figure 1**).

2.1.1.2 Seedling growth and establishment

The first visible structure above the ground is cotyledon. A unifoliate leaf emerges as a true leaf, then the trifoliate, which is the second leaf with three leaflets will be produced. As the growth progress the plant produces alternatively arranged trifoliate leaves. Photosynthesis meets all energy requirements by the alfalfa seedling [29]. The first buds develop in the axils of all leaves, at the point of secondary stem emergence.

2.1.1.3 Contractile growth and crown development

The contractile growth occurs in a process where hypocotyl shortens and has a thicker growth due to carbohydrate storage within plant. This stage pulls the

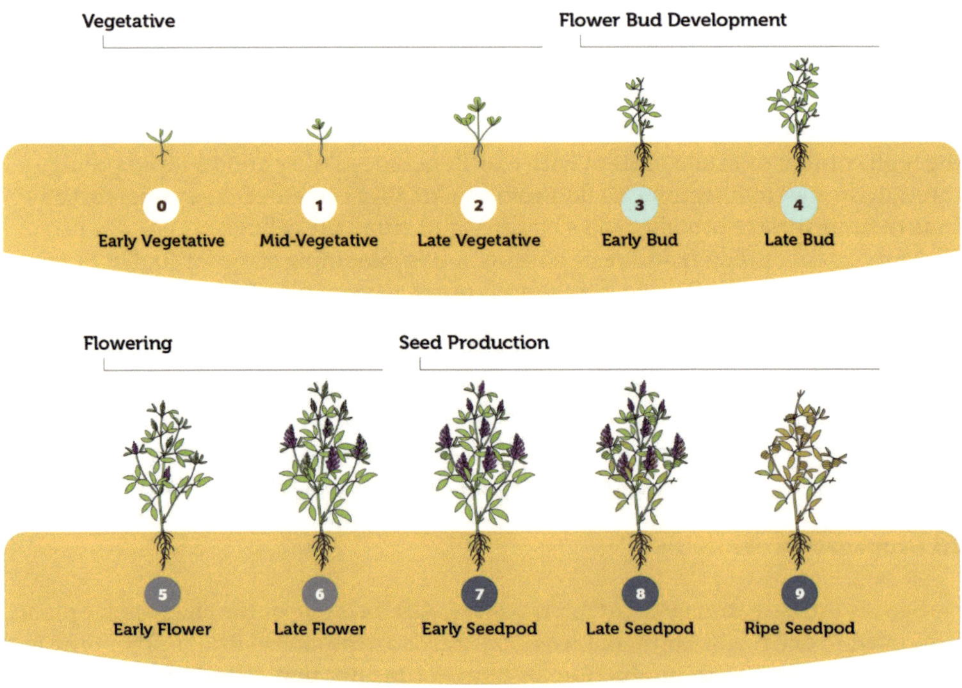

Figure 1.
Growing stages of alfalfa (source: https://nutrien-ekonomics.com/latest-fertilizer-research/alfalfa-development-and-growth-staging)/.

cotyledonary and the unifoliate node underneath and crown develops. At this point of growth, critical crown buds develop. This stage commences from first or second week of emergence and lasts for 2 weeks.

2.1.2 Flowering in alfalfa

The inflorescence of alfalfa is a dense raceme, which is compact around 2 inches long holding up to 50 short-stalked flowers emerging from the leaf axils [30]. The florets commence to open from the base to the tip of the raceme, which takes around a week. A floret may open any hour of the day and remain open for a week if not pollinated, but withers within few hours of pollination (**Figure 2**) [31]. Flowers have five petals, with a larger standard petal upright to curved backward with deep colored veins. The corolla consists of a standard petal, which is the landing platform for pollinators, two smaller wing petals, called lateral petals that project forward and two fused petals called keel. The keel encloses the reproductive parts, two groups of stamen and a style from the green ovary, enclosing anthers and stigma [32]. The color of corolla varies from purple or violet through shades of blue, green, yellow, cream or white [33]. The calyx is short, green with five long pointed lobes with or without fine hair. Each flower has a small pointed bract (**Figure 3**). The sexual organs are non-functional unless released from keel; once released (tripped), fertilization happens if successful pollination occurs. The ovules in the ovary begin to develop, resulting in curly spiral pods twisting in 2 or 3 spirals, the number depends on the number of ovules that progress as complete seed. The number of seeds in a pod depends on degree of pollen compatibility. The pod matures and is equipped for harvest after a month of pollination [34]. A characteristic pod has fine hairs that turns black as matures and release kidney shaped seeds.

2.1.2.1 Tripping

The release of reproductive parts is a phenomenon that is mandatory in alfalfa fertilization and profitable seed production [35, 36]. Tripping is defined as the release of sexual column which includes style, stigma, part of ovary and 10 stamens, by an explosive force as a string under tension is released. Pollinator inserts it proboscis into the flower throat and exerts pressure among the keel petals to access pollen and nectar [37]. When a flower visitor trips the flower, the pistil and stamens hit the pollinators head, where the pollen get deposited and is dislodged by future tripping's of other flowers [38]. Tripping also breaks a membrane on the stigma, which would prevent fertilization of ovule [39]. Natural tripping agents are pollinators such as honey bees and other types of bees and wind [40]. Alfalfa is partially incompatible at fluctuating degrees. If cross-pollinated pollen is not deposited when the flower is tripped, alfalfa may still fertilize by self-pollen (**Figure 4**).

2.1.2.2 Pollination: honey yield, nectar secretion and pollen production

A typical alfalfa plant produces racemes of small pale to dark purple flowers, which require insect visitation for pollination. When a bee visits for pollination, it opens the keel petals and the stamens and pistil move forward, striking the bee with a force. The usual visitors grasp a technique to avoid this mechanism by approaching the flower from the side and inserting their proboscis in between the petals from the base of flower to reach the nectar. Alfalfa secretes a large quantity of nectar, that lures a variety of bee species, of which honey bees produce excellent crop of high-quality honey. The estimated yield of nectar is 416–1933 pounds of nectar per acre [41]. Several terpenoid compounds are identified in alfalfa crops that attracts

Figure 2.
Alfalfa flowers.

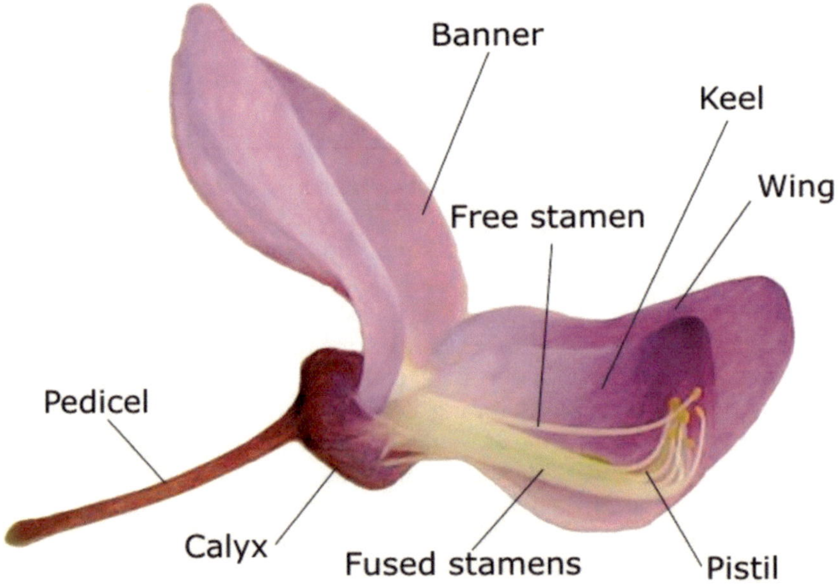

Figure 3.
Inflorescence of alfalfa.

the pollinators towards alfalfa, of which, the aromatic compound ocimene has great role [42, 43]. Alfalfa excretes tremendous quantity of nectar, which lures variety of bees, and from which honey bees produce excellent crops of high-quality honey [41]. Approximately 400–1900 pounds of nectar is secreted by alfalfa [42]. Honey bees are reluctant pollinator of alfalfa as they have to make effort to struck forcefully in the head for the flower to be tripped to enter into forage for nectar. The leaf cutter bee is efficient in pollinating alfalfa than honey bees [44]. High densities

Honey bee in the process of tripping **A tripped flower**

Figure 4.
Tripping.

of bees can lead to adequate pollination to produce desirable honey quantity. A number of 12–15 colonies per acre are recommended for efficient pollination and better honey production. A strong honey bee colony per acre of seed alfalfa could produce 50–100 pounds of honey. Several wild bees also extend their pollination services in alfalfa [45]. Breeding alfalfa cultivars more friendly to honey bees with traits such as plants with stamen protruding beyond keel, plants that could be more easily tripped, plants with high nectar production etc. will aid in greater pollination services, thereby more seed set [46]. Bee colonies in alfalfa crop, yields honey potentially as they are the chief alfalfa pollinators. If the crop is cut for hay before flowering, the bee keeper will get very little quantity of honey [47]. If the crop is for seed production, a better yield of honey can be harvested, which depends on the plant density, competition from other pollinators and other environmental as well as agronomic factors. Alfalfa is a poor source of pollen for honey bees, but relished by other pollinators such as *Bombus*, *Megachile*, *Halictus* and *Nomia* [48].

2.1.2.3 Seed set

A miniature portion of the total ovules develop into seed. A flower should be visited at least once by a pollinator, to get tripped before fertilization. Once tripped, the racemes hold the developing seed pods on the lower part, few open florets in the middle and unopen buds at the top [49]. The cross-pollinated pods produce comparatively more seeds than self-pollinated plants. Moreover, the plants regenerated from self-pollinated plants are less competitive with low survival rates. It is estimated that 2 billion flowers per acre of alfalfa were produced, of which 200 million flowers have the ability to set pods. In a probability of 5 seeds per pod, 2,200,000 seeds per pound, a potential seed production of 50,000 pounds of seed per acre is estimated [50].

2.1.2.4 Seed dispersal

Alfalfa seeds are stood in coiled leguminous pod and it is nonshattered. The local mode of dispersal is natural and non-mechanized. Wind dispersal of seeds is very rare as the seeds are very dense and smooth. Seed dispersal by animals is a possible mode, in situations where animals forage on them [51].

2.1.2.5 Seed quality

Seed quality is the critical aspect in agriculture for offspring generation as well as their conservation, which is measured by germination percentage. The factors influencing seed quality includes proper irrigation and nutrient supply, climatic conditions during seed formation, maturation, harvesting, and drying techniques. Seeding in suitable field condition at proper timing, smearing good crop management aspects, adopting suitable harvesting and drying techniques, minimizing mechanical injuries and ensuring minimum deterioration in storage will favor in obtaining alfalfa seeds of prior quality [52].

2.2 Taxonomy and genetics

Medicago sativa L. belongs to the order Fabales, family Fabaceae, tribe Trifolieae, genus *Medicago*. The genus *Medicago*, is wide, comprises more than 50 species; mostly annuals and few species are perennials [53]. *M. sativa* is a perennial herbaceous legume, that is a tetraploid (2n = 4x = 16) [54]. The allogamy and autotetraploid nature of this species contribute to genetic variations within the population. The commercially cultivated species for forage is *M. sativa* belongs to a *M. sativa* complex, a group of closely related sub species sharing similar karyotypes, the collection of chromosomes and interfertile, capable of interbreeding. The diverse subspecies in the complex comprises subsp. *glutinosa*, subsp. *coerulea* etc. [55]. The most commonly cultivated alfalfa in the globe is *M. sativa* and *M. falcata* is cultivated to some extend around the globe. *M. sativa* is a tetraploid, characterized by purple flowers and coiled pods, whereas *M. falcata* occurs both as tetraploid and diploid and characterized as yellow flowers and sickle shaped. Two other closely related species *M. prostrata* and *M. glomerata* is utilized in natural hybridization with alfalfa [56]. All other members of *M. sativa* species readily cross pollinate with the cultivated species. Several alfalfa species exhibit genetic self-incompatibility or self-sterility and will not successfully self-pollinate [57]. Alfalfa is severely affected by inbreeding depression i.e. reduced forage and seed yield due to self-fertilization attributed to accumulation of recessive alleles due to self-pollination or pollination among close relatives. Alfalfa plays a great role in any genetic engineering context in their improvement as it is an allogamous tetraploid, highly sensitive to inbreeding depression, inability to survive and perpetuate its genetic material [58], highly variable among cultivars and most are synthetic cultivars produced by recurrent phenotypic selection [59]. Therefore, advancement in alfalfa readily lies in other species or sub species of economic importance.

2.3 Ecology of alfalfa

Alfalfa is well adapted to most environments as demonstrated by its wide global distribution. The perennial growth habit of alfalfa, ability to survive extreme temperature, nitrogen fixing symbiosis through their roots and their hard seed coat facilitate their survival in extended ecological regions [60]. High summer temperature, drought, competition from other grasses, water logging and root and crown diseases leads to stand loss [61].

Hard seed: Hard seed is a post-harvest dormancy attributed by alfalfa seeds. A hard water impervious seed coat temporarily prevents water uptake, thereby delays germination. Majority of alfalfa seeds imbibe water and germinate readily and not exhibiting true physiological dormancy [62]. **Auto toxicity:** Alfalfa plants and plant residues produce compounds that provoke an autotoxic reaction that hinder seed germination [63].

2.4 Reproductive ecology of alfalfa

Reproductive success in alfalfa occurs through interaction between development and organization of flower, pollination, fertilization, seed formation, dispersal and other ecological factors. Flowering onset is an intricate progressive evolution that could be activated by various pathways [64]. Diverse ecological and plant internal factors like day length, temperature, light quality and age of plant mediate this transition [65–67]; temperature and photoperiod are important signals in alfalfa flower induction [68, 69]. Alfalfa is a long day plant, requiring 12–14 h of minimal brightness to induce flowering [70]. Alfalfa is solely an insect pollinated crop, which is pollinated only by a small group of insects, especially bees. Each alfalfa bloom is pollinated by a single insect in a single time. The pollinator species, the genotype of alfalfa and the nectar content can affect the pollination process of alfalfa. The predominant pollinators of alfalfa are honey bees (*Apis mellifera*), leaf cutter bees (*Megachile rotunda*) and alkali bees (*Noma melanderi*). Alfalfa seed requires 4–6 weeks to develop into a viable seed [71]. The environmental factors such as rainfall and low temperature during maturing process will decrease the viability of the seeds, ultimately producing seeds of poor quality. The seedling vigor will be reduced and the germination quality will be minimized due to fungal infection or premature seed sprouting [72]. Genetic differences between varieties also affect seed maturity.

3. Advances in improving reproductive ecology of alfalfa

Alfalfa, well known for its extensive ecological adaptability, with development of modern breeding and genetic engineering techniques has advanced in reproductive ecological aspects. Genetic engineering approaches in alfalfa breeding is utilized to manipulate the expression of genes involved in metabolic pathways to improve the reproductive traits, pollination and seed quality traits.

3.1 Breeding and seed production

Any alfalfa cultivar has around 10–200 parents, crossed in isolation to produce breeder seeds. A group of plants distinct for a specific trait, preserved through 2–3 generations from other alfalfa cultivars and cultivars produced by this kind of breeding is called synthetic cultivars. Most commercial alfalfa cultivars are synthetics produced from advanced progenies of clones of superior traits [73]. The out-crossing nature of the alfalfa plant and its polyploid genome create complexity in genetic improvement for higher forage and seed production. The prevalence of severe inbreeding depression prevents researchers from capturing heterosis in alfalfa cultivars through hybrid development. Consequently, it has led to a modification in breeding strategy for higher yield, by intercrossing selected parents to produce synthetic cultivars [74]. This strategy is currently more feasible than the development of a hybrid cultivar. The intercrossing approach between plants with a broad genetic base increase heterozygosity, which increases the intra-locus interaction, and ultimately yield.

3.1.1 Alfalfa, a potential plant for genetic research

The genus *Medicago* is scattered worldwide and comprises 83 species [75]. The cultivated alfalfa for fodder purpose is autotetraploid, cross-pollinated and seed propagated [76]. A rich source of natural variation and genetic resources is held by three species *M. sativa*, *M. falcata* and *M. media*, with breeding potential for development of new cultivars. Genetic diversity studies of any species are indispensable for successful breeding into new cultivars. Variations of a population, genetically and phenotypically give insights in evolutionary history [77]. A study with 25 accessions of the three *Medicago* species collected in Leh Region, Kashmir, India, analyzed using simple sequence Repeats (ISSRs) and random amplified polymorphic DNA revealed a high genetic variation. A genetic diversity of 30.23% within populations with a mean coefficient of differentiation (Gst) was noticed. The overall value of mean estimated number of gene flow (Nm = 8.0682) revealed large gene exchanges among populations. Analysis of molecular variance (AMOVA) states the genetic diversity being 49% among populations and 51% between populations. The distinct genetic variation in *Medicago* species is an ideal plant for genetic research for breeding for improved forage varieties.

3.1.2 Breeding to overcome inbreeding expression

Several breeding programs are structured to avoid or overlook the negative impacts of inbreeding expression in alfalfa [72]. Inbreeding expression occurs due to continuous selfing. The vigor of natural tetraploid alfalfa is reduced by inbreeding depression, but maximizing heterozygosity offers a better solution [92]. The intra-locus interaction and additive variation maximizes heterogeneity in alfalfa, to enhance its performance [78]. The forage yield of alfalfa is experienced due to increased frequency of favorable alleles and utilization of non-genetic effects in a study for long years (1898–1985). The study also revealed the enhanced genetic load in recent alfalfa cultivars due to several crossings, which hinders the lethal alleles at their heterozygous loci, providing better germplasm source of alfalfa [79].

3.2 Advances in floral biology

3.2.1 Effect of anatomical structure of the alfalfa flower in tripping mechanism

Alfalfa possess a highly unique papilionaceous flower with an exclusive tripping mechanism, which permits only certain insects to enter in to effect tripping [94]. *M. sativa* was studied at the blossom stage to correlate the part of flower tripped by pollinators in relation to the blossom characters. The comprehensive study on the anatomical structure of flowers of 11 selections displayed two chief services involved in tripping mechanisms *viz.*, a cohesive force between the two keel petals and the pressure exerted by the sexual column from cells under tension at the point of fused filaments and the keel. The initial force was resulted due to the interlocking of the projections on the pressed surfaces of keel petals, which is adequate to prevent spontaneous tripping of alfalfa flowers. The firmness of the preventive force and the relative pressure exerted by the sexual column against the appressed keel petals differs in flowers from various selections which is attributed to differences in anatomical structure and development of blossoms. The results revealed that the preventive force of the adhered keel petals is closely related to the proportion of insect tripped flowers than the force with which the sexual column is released by tripping [80].

3.2.2 Mutants of varied inflorescence patterns in alfalfa

Alfalfa is a typical leguminous plant with a typical raceme inflorescence. The multi-inbreeding process in this cross-pollinated species, develop several mutant forms with long, panicle-like racemes, with fertile and sterile flowers, complex branched racemes, fascinated racemes etc. The transitional form of some mutants is isolated by means of pair hybridization and new mutant forms were developed. New mutants with diverse inflorescence pattern provide room for more pollinator's activity leading to improved seed set. *Medicago trunculata* shared highly conserved nucleotide sequence and exhibit perfect synchrony between genomes. *M. trunculata* mutant *mtpim* has a complicated inflorescence resembling panicle, controlled by spatiotemporal expression of MtTFL1, MtFULC, MtAP1 and SG1 through reciprocal repression. Some of them resemble *M. sativa* phenotypes. The mutant developed by retrotransposon insertion mutagenesis sg1l-1 has a cauliflower type phenotype resembling a mutant of alfalfa [81]. The data generated on genes regulating inflorescence developed in *Medicago* species helps to understand the phenomenon of inflorescence mutations in alfalfa, which is helpful for modification in inflorescence structure for enhanced pollination.

3.2.3 Ovule sterility and seed set in alfalfa

The seed potential of alfalfa is very low; an estimated seed to ovule ratio is about 0.08 [82]. This deficit is due to the low number of seeds produced per pod; of 10 ovules present in a floret, only an average of 5 develops into seed. Eliminating at least few causes that limit seed potential will be beneficial in improving seed yield of alfalfa. An ovule sterility trait allied with limiting integument formation controlled by a single recessive gene has been developed in alfalfa. In alfalfa cultivar, Blazer XL, an ovule sterility trait B17, with 81% ovules displaying heavy callose deposition at the time of anthesis with low female fertility was reported [83]. These plants were female sterile when hand crossed with unrelated male fertile plant. In high percentage of ovule sterile plants, under sized pistils develop at anthesis, that will not emerge from staminal column. A mapping study of chromosome region revealed and explains a major share of variation for ovule sterility and the cytological analyses showed that no embryo sac develops in sterile ovules and the callose deposition begins after meiotic division [84], affecting the integumentary tapetum and nuclear cell walls, which usually expands to fill the space occupied by the embryo sac [85]. The pistil growth often ceases at bud stage and a short-aborted pistil is found within the staminal column at anthesis, in a plant with 100% sterile ovules [86]. Nine populations analyzed for ovule sterility showed 4–26% sterile ovules with significant negative correlation between percentage of sterile ovules and seeds per pod in most populations [87]. A quick stain clearing technique based on callose fluorescence is effective to trace ovule sterility in breeding programs. In addition, checking ovule fertility in parental genotypes aid breeders to develop good seed yielding cultivars.

3.2.4 Effect of floral nectaries and flower aroma of alfalfa

Alfalfa secrets nectar at a uniform daily rate for 4–5 days after flower opening. The volume of the nectar produced differs per floret, which are found to be heritable. Two plants each from a high, intermediate and low nectar producing alfalfa cultivars were subjected to light and scanning electron microscopy, and the images designated nectary located on the staminal column of the receptacle [88]. It comprises of several cell layers subtended by vascular bundles containing both xylem and phloem, but not extend into nectariferous tissue. The epidermal layer comprises permanently open stomata, which functions in nectar secretion. The number of stomata per nectary among the six clones ranged from 24.7 ± 1.9 to 6.8 ± 0.5. The nectar-reservoir diameters ranged from 1.07 ± 0.09 mm to 0.70 ± 0.01 mm. The cultivars with the largest nectar reservoir witnessed high number of stomata [89].

Flower aroma of alfalfa has a specific role in pollinator attraction. Seven clonal lines of alfalfa with difference in flower aroma were consecutively recorded for honey bee visits at different locality and volatiles were also extracted from the flowers. A gas chromatographic study suggests a difference in volatile components [90]. Therefore, the compounds present in the volatiles of nectar has a major role in pollinator attraction for enhanced seed set.

3.2.5 Features of alfalfa flower that effect seed production

The characters and properties of alfalfa flowers investigated to determine their role in seed setting revealed that the affluence to trip the flower is allied with the age of floret. The age of floret does not have an effect in self-fertilization, but declines for 3 days in cross fertilization. Pollen from untripped florets showed decline in germination, but remained viable when stored in honey bee colony. The honey bees that collect pollen clean themselves with pollen in 2 days. There was no steadfast increase in seed production due to repeated visits of honey bees. Dusting of foreign pollen before tripping has not increased cross fertilization, but 50% increase in self-fertilization. Ovules of florets with exposed stigmas remained functional up to 6 h and then declined rapidly for next 2 days. Self and cross-fertilized seeds were amalgamated in the pods but self-fertilized seeds did not occur beyond the fourth position from the proximal end. Studies suggested that the honey bee pollination can be improved by providing an alfalfa flower with an exposed stigma [91].

3.3 Influence of pollinators in alfalfa pollination and potential for gene flow

Diverse pollinators have diverse roles in pollinating specific crops. Though managed pollinators, are utilized for pollination of alfalfa for seed production, several wild pollinators also visit them [92–95]. Diverse pollinators display assorted efficiency in depositing and detaching pollen from individual flowers [96, 97]. The rate of tripping and the quantity of pollen deposited varies between pollinator species visiting the racemes [98], which depends on whether the pollinator forages for pollen or for nectar [99] and influence on seed set. Apart from influencing pollination, the pollinators differentially affect gene flow [100]. The bumble bees carry pollen for short distances, but not the hawkmoths. In addition to pollinators, various features of field that grow alfalfa also plays a major role. Increased density of plants has displayed reduced gene flow as pollinators respond to locally available floral resources and shorten the flight distance [101–103]. The load of pollen carried by the pollinators from donor to recipient is expected to turn over quickly as pollinator visits greater number of plants per unit distance traveled, which declines the gene flow.

Though transgenic alfalfa plants are developed, the impact of pollinators on gene flow is yet to be studied as various wild insect pollinators contribute to pollination and movement of genes [104]. A model to predict gene flow between transgenic and conventional field [105], predicted the number of visits by the pollinator in a foraging session, estimates the extent to which the transgenic pollen is diluted by conventional pollen. Essentially, the amount of transgenic pollen on conventional field flower is inversely proportional to the total amount of pollen delivered by each bee during a foraging session in conventional alfalfa. Accordingly, the greater number of flowers the specific pollinator visits, the minimal amount of transmission of transgenic pollen and its fruit set [106]. A study to investigate the involvement of managed honey bees (*A. mellifera*) and leafcutter bees (*Megachile rotundata*), three bumble bee species (*Bombus impatiens*, *Bombus griseocollis*, *Bombus auricomus*) and two solitary bee species to pollination (*Halictus rubicundus* and *Andrena asteris*) and their contribution in gene flow revealed the potential role of two wild solitary bees and a wild bumble bee in tripping, while the managed pollinators, *A. mellifera* and *M. rotundata* recorded the least. Honey bees, recorded the best potential in gene flow and reduced transgene transmission in relation to other pollinators. The denser plant stand of alfalfa does not show any impact in gene flow and reduced transgene transmission, while the three bumble bee species portrayed an increased gene flow and reduced transgene escape in high density alfalfa plantations [107].

3.3.1 Gene flow in commercial seed production

Alfalfa being an outcrossing and insect pollinated crop, the potential for gene flow has been widely recognized. Gene flow is the exchange of genes from one population to another. It is a natural mechanism that changes the genetic frequency of population over time to enrich the wealth of biological diversity. The formation of cross-pollinated genes and their establishment mediates occurrence of real gene flow [108]. Though pollinator species carry pollen to long distances, true gene flow can occur if the pollen can produce viable seeds and thus offspring. Pollen mediated gene flow is greatly dependent on physical isolation distance between two populations, the degree of synchronous flowering, availability of pollinator gene frequency etc. [109]. Maintaining adequate isolation distances between any alfalfa and seed production trials help to minimize the bee-mediated gene movements. A minimum of 50 m between the alfalfa crops for certified seed production and 200 m for plots exceeding 5 acres and 300 m for plots of 5 acres or less for foundation seed production is recommended [110, 111]. In seed alfalfa production for commercial purpose, bees are purposely stocked in the field. Gene flow, rate of outcrossing, pollinator type and behavior are very important for proper management of commercial alfalfa seed production. Advanced genetic technologies offer better tools to understand the dynamics of pollinator-mediated gene flow. A study to measure the gene flow using honey bee pollination under commercial alfalfa seed production reported a significant decrease in gene flow with increase in isolation distance. A 900 ft. colonies of honey-bee (*A. mellifera*) mediated gene flow was 1.49% and it decreased linearly to 0.20% near 5000 ft. Gene flow continue to decline with increase in isolation distance [112]. The same was reported for alfalfa leaf cutter bee (*M. rotundata*) [113]. Another research with Roundup Ready Trait as a marker gene flow examined the movement of the gene *epsps* from the Roundup Ready plots to conventional alfalfa plots isolated by 152, 274, 610 and 825 m distance. With leaf cutter bee as the pollinator, the pooled data from 2000 to 2002 revealed that the upper bound (99.9% confidence) of gene flow at 274 m was 0.3%, at 152 m was 1.7% and gene flow was not detected at 610 and 825 m at years 2000 and 2002. The results confirmed the role of isolation distance as an effective means to maintain purity of both Roundup Ready and conventional alfalfa plots [114].

3.3.2 Gene flow in forage alfalfa production

Pollen flow or dispersal is a form where genes can move between plants. However, the total pollen dispersed will not cause gene flow, which is the successful transfer of genetic material. As pollen-mediated gene flow occurs between sexually compatible plants and deposited on stigma of a plant, it fertilizes the ovule, resulting in production of viable seed. Flowering in alfalfa reduces the forage and nutritive value and it has significant consideration in gene flow of alfalfa. Alfalfa managed for forage production is cut several times a year from two to ten, depending upon the climatic factors of the region and the stage of alfalfa development which is early to 10% blooming stage. Periodical harvesting removes the entire plant canopy with flowers or seeds. For high quality forage, first cutting occurs in mid to late bud stage [115]. The cutting interval is 28–35 days, an interval well adequate to initiate full bloom and matured seed. Harvest of alfalfa as dairy feed eliminated the entire plant canopy terminating bloom and seed formation. Regrowth of canopy reinitiated from vegetative crown buds and elongation of lower stem axillary buds. Therefore, alfalfa managed for forage production will have minimal contribution to gene flow, as flower blooming happens in a very less percent and thus pollen transfer. Gene flow from Roundup trait alfalfa to conventional alfalfa for forage production is predicted to be far less as compared to seed production alfalfa [116]. Accordingly, the gene flow from genetic engineered alfalfa grown for forage to conventional forage field is lower than alfalfa grown for seed.

4. Climate change

Alfalfa, extends its timely service in climate regulation by drawing carbon from atmosphere and dumping deep in soil. But climate change effected by global warming and enhanced nitrogen deposition directly or indirectly affect the reproductive characters of alfalfa. The shift in phenology and foraging crop distribution of plants and pollinators leads to temporal decoupling and spatial mismatch between them [117]. Climate change may affect the floral traits such as floral display, corolla structure, formation and position of ovule and stamen, subsequently affecting the quality and quantity of forage of pollinators, ultimately the reproductive success in plants. Insect pollinators are valuable resources but are limited. Although around 20,000–30,000 bee species are present worldwide, only 10–11 species are managed [118], and *A. mellifera* is far being the most dominant species managed globally. Biotic stress convoyed with climate change causes severe decline in pollinator population. The genus *Apis* is in risk due to climate change and the mismatch in foraging plants and pollinator regimes, thus there is a want for alternate pollinators. Well-known pollinator, *A. mellifera* is being replaced by leaf cutter bee (*M. rotundata*) and alkali bees (*Nomia melanderi*) in alfalfa ecosystem [119].

5. Future scientific challenges and perspectives

Alfalfa cultivars are usually synthetic populations that originated by a heterogenous blend of heterozygous genotypes, that thwarts the genomic results in breeding process. The application and implementation of genomic techniques for genetically improved alfalfa may be quite promising and challenging, as most matters could be solved at technical and commercial level. Even though, plant genetic engineering encounters the chances for improvement of alfalfa cultivars to some extent, but there are still challenges [120]. Gene transfer techniques involve appropriate use of promoters, transit peptides, choice of selectable or reporter markers etc. The promotors specific to alfalfa are inadequate and fully characterized expressing genes in high level. Consequently, a convention of constitutive, specific promoters effective to alfalfa are mandatory for augmented expression of transgenic research. Therefore, a set of constitutive, tissue or temporal-specific promoters effective in alfalfa is obligatory for the optimized expression of transgenic research [121]. As alfalfa is tetraploid, the transgene integration and gene stacking techniques utilized for other crops are not suitable for alfalfa. Innovative breeding approaches would be needed to adopt and address these challenges in alfalfa.

6. Conclusion

Reproduction is a crucial phase in the life period of any organisms, which ensures their endurance as a distinct species and governing the adaptability of plants to multivariate ecosystem. The precise design of resource distribution for reproduction depends on the reproductive parts and their union, factors influencing the fertilization, genetic variation and resource availability. Plants choose diverse reproductive modes based on the resource and pollination deficiency. Self-pollination and clonal reproduction are conducive to reproductive assurance in stands with pollinator scarcity, while sexual reproduction and cross-pollination improves seed quality and genetic diversity of progenies, which would benefit population survival. Consequently, the amalgamations of assorted reproductive modes and generalized pollinator networks may be good choices to adapt successful reproduction. Awareness on the key processes can be useful in addressing the current challenges, that impact the ecosystem directly or indirectly. The evolution of floral traits might be obsessed by selective pressure of local pollinators, exotic pollinators, nectar robbers and abiotic environments such as temperature, precipitation and ultraviolet radiation on several traits related to flower. The applications of new technologies and methods such as breeding to overcome inbreeding depression could advance the awareness on alfalfa reproductive ecology. Advanced floral mechanisms such as altered anatomical structure of the alfalfa flower in tripping mechanism, mutants of varied inflorescence patterns, ovule sterility and seed set, floral nectaries, floral aroma and flower features of alfalfa has major influence on altering its reproductive ecology. Outstandingly, the progress in high-throughput sequencing of alfalfa cultivars genetically, has brought an outlet for the various genetic shortcomings in alfalfa mass production accomplished by breeding and biotechnological means. The fundamental mechanisms on diverse topics of plant reproductive ecology could be discovered more effortlessly, especially in relation to floral traits and pollinators for better alfalfa production in future researches.

Author details

Hana'a Burezq
Desert Agriculture and Ecosystems Program, Environment and Life Sciences Research Center, Kuwait Institute for Scientific Research, Safat, Kuwait

*Address all correspondence to: haborizq@kisr.edu.kw

References

[1] Tandon R, Shivanna KR, Koul M. Reproductive ecology of flowering plants: An introduction. In: Reproductive Ecology of Flowering Plants. United States: Springer Publication; 2020. p. 1

[2] Armbruster WS, Mulder CP, Baldwin BG, Kalisz S, Wessa B, Nute H. Comparative analysis of late floral development and mating-system evolution in tribe Collinsieae (Scrophulariaceae sl). American Journal of Botany. 2002;**89**:37-49

[3] Kruckeberg AR, Rabinowitz D. Biological aspects of endemism in higher plants. Annual Review of Ecology and Systematics. 1985;**16**: 447-479

[4] Morely T. Flowering frequency and vegetative reproduction in *Erythronium albidum* and *E. propullans,* and related observations. Bulletin of the Torrey Botanical Club. 1982;**109**:169-176

[5] Karron JD. The pollination ecology of co-occurring geographically restricted and widespread species of *Astragalus* (Fabaceae). Biological Conservation. 1987;**39**:179-193

[6] Deyrup M, Menges ES. Pollination ecology of the rare scrub mint *Dicerandra frutescens* (Lamiaceae). Florida Scientist. 1997;**60**:143-157

[7] Bell G. On the function of flowers. Proceedings of the Royal Society of London B. 1985;**224**:223-265. DOI: 10.1098/rspb.1985.0031

[8] Ashman TL. Pollen limitation of plant reproduction: Ecological and evolutionary causes and consequences. Ecology. 2004;**85**:2408-2421. DOI: 10.1890/03- 8024

[9] Aguilar R, Ashworth L, Galetto L, Aizen MA. Plant reproductive susceptibility to habitat fragmentation: A review and synthesis through a meta-analysis. Ecology Letters. 2006;**9**:968-980. DOI: 10.1111/j. 1461-0248.2006.00927.x

[10] Barrett SCH. Mating system evolution and speciation in heterostylous plants. In: Otte D, Endler J, editors. Speciation and Its Consequences. Sunderland, MA: Sinauer; 1989. pp. 257-283

[11] Barrett SCH, Harder LD, Worley A. The comparative biology of pollination and mating. Philosophical Transactions of the Royal Society of London. Series B, Biological Sciences. 1996;**351**: 1271-1280. DOI: 10.1098/ rstb.1996.0110

[12] Willson MF. Sexual selection in plants. The American Naturalist. 1997;**113**:777-790. DOI: 10.1086/283437

[13] Morgan MT, Schoen DJ. The role of theory in an emerging new plant reproductive biology. Trends in Ecology & Evolution. 1997;**12**:231-234. DOI: 10.1016/S0169-5347(97) 01045-8

[14] Harder LD, Barrett SCH. Ecology and Evolution of Flowers. Oxford, UK: Oxford University Press; 2006

[15] Radović J, Sokolović D, Marković J. Alfalfa most important perennial forage legume in animal husbandry. Biotechnology in Animal Husbandry. 2009;**25**(5-6):465-475

[16] Putnam D, Ruselle M, Orloff S, Kuhn J, Fitzhugh L, Godfrey L, et al. Alfalfa Wildlife and the Environment. The Importance and Benefits of Alfalfa in the 21st Century. Novato, CA: The California Alfalfa and Forage Association; 2001. p. 24

[17] Michaud R, Lehman EF, Rambaugh MD. World distribution historical development. In: Alfalfa and

Alfalfa Improvement. Madison, Wisconsin, USA: A.A. Hanson, D.K Barnes an Publishers; 1988. p. 1084

[18] ITC. International Trade Center. 2018. Available from: http://www.intracen.org/itc/market-infotools/trade-statistics

[19] Mishra SB, Singh UK, Sadu PK. Studies on genetic divergence in five land races of rice in northern Bihar. Annals of Agriculture Research. 1994;**15**(2):217-221

[20] Dinic B, Dordevic N, Radovic J, Ignjatovic S. Modern procedures in technology of conserving lucerne in ensiling. Biotechnology in Animal Husbandry. 2005;**21**:5-6, 297-303

[21] Derscheid LA, Walstrom RJ. Alfalfa Seed Production. Extension Circulars. Paper 433. 1981. Available from: http://openprairie.sdstate.edu/extension_circ/433

[22] Berdahl JD, Wilton AC, Frank AB. Survival and agronomic performance of 25 alfalfa cultivars and strains inter-seeded into rangeland. Journal of Range Management. 1989;**42**:312-316

[23] Geng HZ. Evaluation and selection of the salt-tolerant alfalfa. In: Proceedings of the SVI International Grassland Congress, Nice, France. Leach, G.J. and Clements, R.J. 1984. Ecology and grazing management of alfalfa pastures in the subtropics. In: Brady NC, editor. Advances in Agronomy. New York, USA: Academic Press Inc.; 1989. pp. 127-154

[24] Muller MH, Ronfort J, Santoni S, Prosperi JM. Consequence of domestication bottleneck on the DNA sequence diversity of a nuclear gene in wild and cultivated *Medicago sativa*. In: Proceedings of the North American Alfalfa Improvement Conferences (NAAIC). 2002. pp. 55-65

[25] Patnam DH. Will the public recognize the environmental benefits of alfalfa? In: Proceedings of the 2004 National Alfalfa Symposium, San Diego, CA, USA. 2004

[26] Iqbal S, Mushtaq A, Gul Z, Faizan A, Eajaz AD. Alfalfa (*Medicago sativa* L): A valuable source of forage/fodder crop under cold arid conditions of Ladakh (Jammu and Kashmir). Research Journal of Chemical and Environmental Sciences. 2018;**6**(4):74-84

[27] Koochekin A, Marashi H. The effect of plant spacing and cutting treatments on seed yield of alfalfa. Mashhad, Journal of Agricultural Science and Technology. 1989;**3**(3):50-64 (In Persian)

[28] Rashmi R, Sarkar M, Vikramadithya. Cultivation of Alfalfa (*Medicago sativa*). Ancient Science of Life. 1997;**17**(2):117-119

[29] Uphof JCTH. Dictionary of Economic Plants. 2nd ed. New York: Verlag Von J. Caramer; 1968. p. 335

[30] Larkin RA, Graumann O. Anatomical structure of the alfalfa flower and an explanation of the tripping mechanism. International Journal of Plant Sciences. 1954;**116**:116-123. DOI: 10.1086/335845

[31] Nielsen HM. Floral Modification in Lucerne. In: 1st Internatl. Symposium on Pollination Proc., Copenhagen, Aug. 1960. Commun. 7, Swedish Seed Growers' Assoc. Agricultural Handbook. pp. 60-63

[32] Ggraumann HO, Hhanson CH. Growing Alfalfa. Vol. 1722. United States: Dept. Agr. Farmers' Bul; 1954. p. 38

[33] Barnes DK. A System for Visually Classifying Alfalfa Flower Color. Vol. 424. United States: Dept. Agr., Agr. Handb.; 1972. p. 12

[34] Pankiw P, Bolton L. Characteristics of alfalfa flowers and their effects on seed production. Canadian Journal of Plant Science. 1965;**45**(228): 333-342

[35] Henslow C. Note on the structure of *Medicago sativa* as apparently affording facilities for the intercrossing of sistinct flowers. Journal of the Linnean Society. 1867;**9**:327-329

[36] Cockerell TDA. Some insect pests of the salt river valley and remedies for them. Arizona Agricultural Experiment Station Bulletin. 1899;**31**:273-295

[37] Brown DE, Bingham ET. Selfing in an alfalfa seed production field. Crop Science. 1994;**34**:1110-1112

[38] Tysdal HM. Is tripping necessary for seed setting in Lucern strains? American Society of Agronomy Journal. 1940;**38**: 515-535

[39] Zaleski. Lucerne investigation II. Pollination and seed setting in Lucerne Strains. Journal of Agricultural Science. 1956;**48**:236-44

[40] Brunet J, Stewart CM. Impact of bee species and plant density on alfalfa pollination and potential for gene flow. Psyche. 2010;**3**:98-107. DOI: 10.1155/2010/201858

[41] Kraopcova. Nectar production of Lucerne and the number of honey bees working on it. Sborn. Vysoke Skoly Zemedelske v Brne. 1963;**1**:37-44. [In Czech., Russian and English summaries.]

[42] Loper GM. *Medicago sativa* and *Citrus depressa* flower volatiles. Phytochemistry. 1972;**11**:1865

[43] Fischer RL. Native *pollinators* of alfalfa *Medicago sativa* L. in northern Minnesota. Minnesota Beekeeper. 1953;**6**:8-9

[44] Pitts-Singer TL, Bosch J. Nest establishment, pollination efficiency, and reproductive success of *Megachile rotundata* (Hymenoptera: Megachilidae) in relation to resource availability in field enclosures. Environmental Entomology. 2010;**39**:149-158

[45] Free JB. Insect Pollination of Crops. 2nd ed. London: Academic Press; 1993

[46] Scott-Dupree CD, Winston M, Hergert G, Jay SC, Nelson D, Gates J, et al. A Guide to Managing Bees for Crop Pollination. Aylesford NS: Canadian Association of Professional Apiculturists; 1995

[47] Kevan PG. Pollination, Crops and Bees. Mann, L.K.: OMAFRA Publication; 1988. p. 72

[48] Bohart GE. Pollination of alfalfa and red clover. Annual Review of Entomology. 1957;**2**:355-380

[49] Whornham G. Alfalfa seed investigations and other crop varietal studies Millaed county, Utah, 1929-33 (Inclusive). Utah Agricultural Experiment Station Bulletin. 1936; **268**:24

[50] Lesins K. Investigations into seed setting of lLucerne at Ultuna Sweden. 1945-1949. LESINS. Uppsala, Sweden: Annals of Royal Agricultural College. 1950;**17**:441-483

[51] Blackshaw RE, Rode LM. Effect of ensiling and rumen digestion by cattle on weed seed viability. Weed Science. 1991;**39**:104-108

[52] Rao K, Dulloo ME, Engels JMM. A review of factors that influence the production of quality seed for long-term conservation in genebanks. Genetic Resources and Crop Evolution. 2017;**64**:1061-1074. DOI: 10.1007/s10722-016-0425-9

[53] Quiros CF, Bauchan GR. The genus Medicago and the origin of the *Medicago*

sativa complex. In: Hanson AA, Barnes DK, Hill RR Jr, editors. Alfalfa and Alfalfa Improvement. Madison, WI: ASA-CSSA-SSSA; 1988. pp. 93-124

[54] Riday H, Johnson D, Heyduk K, Raasch J, Darling M, Sandman J. Paternity testing in an autotetraploid alfalfa breeding polycross. Euphytica. 2013;**194**:194-199.. DOI: 10.1007/s10681-013-0938-7

[55] Viands DR, Sun P, Barnes DK. Pollination control: Mechanical and sterility. In: Hanson AA, Barnes DK, Hill RR Jr, editors. Alfalfa and Alfalfa Improvement. Madison, WI: ASA-CSSA-SSSA; 1988. pp. 931-960

[56] Rumbaugh MD, Caddel JL, Rowe DE. Breeding and quantitative genetics. In: Hanson, A.A. (ed.) Alfalfa and Alfalfa Improvement. Agronomy. 1988;**29**:777-805

[57] Stanford EH. Tetrasomic inheritance in alfalfa. Agronomy Journal. 1951;**43**: 222-225

[58] Armstrong JM. Cytological studies in alfalfa polyploids. Canadian Journal of Botany. 1954;**32**:531-542

[59] Flajoulot S, Ronfort J, Baudouin P, Barre P, Huguet T, Huyghe C, et al. Genetic diversity among alfalfa (*Medicago sativa*) cultivars coming from a breeding program, using SSR markers. Theoretical and Applied Genetics. 2005;**111**:1420-1429

[60] McKenzie JS, Paquin R, Duke SH. Cold and heat tolerance. In: Hanson AA, Barnes DK, Hill RR, editors. Alfalfa and Alfalfa Improvement. Madison, WI: ASA-CSSA-SSSA; 1988. pp. 259-302

[61] Leach GJ, Clements RJ. Ecology and Grazing Management of Alfalfa Pastures in the Subtropics. Advances in Agronomy. 1984;**37**:127-154

[62] Undersander D, Albrect K, Degenhart N, Moutray V, McCaslin M. Hard Seed in Alfalfa. University of Wisconsin Extension; 1993. Available from: http://www.uwex.edu/ces/forage/pubs/hardsd.htm [Accessed: 11 June 2008]

[63] Canevari M, Putnam DH, Lanini WT, Long RF, Orloff SB, Reed BA, Vargas RV. Overseeding and Companion Cropping in Alfalfa. University of California Agriculture and Natural Resources Publication No. 21594. California Alfalfa Workgroup; 31 pp. Available from: http://anrcatalog.ucdavis.edu/Alfalfa/21594.aspx [Accessed: 11 June 2008]

[64] Amasino R. Seasonal and developmental timing of flowering. The Plant Journal: For Cell and Molecular Biology. 2010;**61**:1001-1013

[65] Cerdan PD, Chory J. Regulation of flowering time by light quality. Nature. 2017;**423**:881-885

[66] Hyun Y, Richter R, Coupland G. Competence to flower: Age controlled sensitivity to environmental cues. Plant Physiology. 2017;**173**:36-46

[67] Lorenzo CD, Sanchez-Lamas M, Antonietti MS, Cerdan PD. Emerging hubs in plant light and temperature signaling. Photochemistry and Photobiology. 2016;**92**:3-13

[68] Nittler LW, Gibbs GH. The response of alfalfa varieties to photoperiod, color of light and temperature 1. Agronomy Journal. 1959;**51**:727-730

[69] Pearson CJ, Hunt LA. Effects of temperature on primary growth of alfalfa. Canadian Journal of Plant Science. 1972;**52**:1007-1015

[70] Major DJ, Hanna MR, Beasley BW. Photoperiod response characteristics of alfalfa (*Medicago sativa* L.) cultivars. Canadian Journal of Plant Science. 1991;**71**:87-93

[71] McGregor SE. Insect Pollination of Cultivated Crop Plants. Washington, D.C.: United States Department of Agriculture, Agriculture Handbook No. 496; 1976

[72] Rincker CM, Marble VL, Brown DE, Johansen CA. Seed production practices. In: Hanson AA, Barnes DK, Hill RR, editors. Alfalfa and Alfalfa Improvement (Agronomy *Monograph* No. 29). Madison, Wisconsin: ASA–CSSA–SSSA; 1988. pp. 985-1021

[73] Hanson AA, Barnes DK, Hill RR. Alfalfa and Alfalfa Improvement-Agronomy Monograph no. 29. American Society of Agronomy; 1988

[74] Hill RR. Principles of Cultivar Development. In: Fehr W, editor. Principles of Cultivar Preprints (www.preprints.org) | NOT PEER-REVIEWED | Posted: 13 November 2020. Development. Theory and Technique. Vol. 1. Stuttgart, Germany: Macmillan; 1987. Doi: 10.20944/preprints202011.0390.v1 16 636

[75] Small E, Jomphe M. A synopsis of the genus *Medicago* (Leguminosae). Canadian Journal of Botany. 1989;**67**: 3260-3294

[76] McCoy TJ, Bingham ET. Cytology and cytogenetics of alfalfa. In Alfalfa and alfalfa improvement. In: Hanson AA, Barnes DK, Hill RR, editors. Agronomy Monograph 29. Madison, USA: American Society of Agronomy; 1988. pp. 737-776

[77] Barrett SC, Harder LD, Worley AC. The comparative biology of pollination and mating in flowering plants. Philosophical Transactions of the Royal Society of London Series. 1996;**351**: 1271-1280

[78] Dunbier MW, Bingham ET. Maximum Heterozygosity in Alfalfa: Results Using 605 Haploid-derived Autotetraploids. Crop Science. 1975;**15**(4):527-531, 606. DOI: 10.2135/cropsci1975.0011183X001500040023xa

[79] Holland JB, Bingham ET. Genetic improvement for yield and fertility of alfalfa 638 cultivars representing different eras of breeding. Crop Science. 1994;**34**(4):953-957, 639. DOI: 10.2135/cropsci1994.0011183X003400040022x

[80] Barnes DK, Bingham ET, Axtell JD, Davis WH, Hanson CH. The flower, sterility mechanisms, and pollination control. Agronomy Monographs. 1976;**5**:67-72. DOI: 10.2134/agronmonogr15.c6

[81] Dzyubenko NI, Dzyubenko EA. Mutants of inflorescence development in alfalfa (*Medicago sativa*). Journal of Genetics and Breeding. 2019;**23**(6): 700-707

[82] Lorenzetti F. Achieving potential herbage seed yield in species of temperate regions. In: Proc. XVII. International Grassland Congress. 1993. pp. 1621-1628

[83] Rosellini D, Lorenzetti F, Bingham ET. Quantitative ovule sterility in Medicago sativa. Theoretical and Applied Genetics. 1998;**97**:1289-1295

[84] Tiwari SC. Callose in the walls of mature embryo sac of Torenia fournieri. Protoplasma. 1982;**110**:1-4

[85] Barcaccia G, Mazzucato A, Falcinelli M, Veronesi F. Callose localization in cell walls during meiotic and apomeiotic megasporogenesis in diploid alfalfa. Caryologia. 1996;**49**: 45-56

[86] Pereira TNS, Ilarslan H, Palmer RG. Genetic and cytological analyses of three lethal ovule mutants in soybean (*Glycine max*; Leguminosae). Genome. 1997;**40**:273-285

[87] Dzyubenko NI, Vyshniakova MA. Intraspecific variability of seed

productivity character in alfalfa. In: Proc. Third International Herbage Seed Conference, Halle, Germany, June 23-28. 1995

[88] Teuber LR, Albertsen MC, Barnes DK, Heichel GH. Floral nectaries of Alfalfa. American Journal of Botany. 1980;**67**(4):433-439. DOI: 10.2307/2442282

[89] Loper GM, Waller GD. Alfalfa Flower Aroma and Flower Selection by Honey Bees. Joint contribution from Crops Research Division and Entomology Research Division, Agricultural Research Service, U. S. Department of Agriculture. Arizona Agric. Experiment Sta., 1970. DOI: https://doi.org/10.2135/cropsci1970.0011183X001000010026x.

[90] Pankjw P, Bolton JL. Characteristics of alfalfa flowers and their effects on seed production. Canadian Journal of Plant Science. 1965;**3**:41-45. DOI: 10.4141/cjps65-066

[91] Bohart GE. Pollination of alfalfa and red clover. Annual Review of Entomology. 1957;**2**:355-380

[92] Cane JH, Schiffhauer D. Dose-response relationships between pollination and fruiting refine pollinator comparisons for cranberry (*Vaccinium macrocarpon* [Ericaceae]). American Journal of Botany. 2003;**90**(10): 1425-1432

[93] Hayter KE, Cresswell JE. The influence of pollinator abundance on the dynamics and efficiency of pollination in agricultural *Brassica napus*: Implications for landscape-scale gene dispersal. Journal of Applied Ecology. 2006;**43**(6):1196-1202

[94] Cecen S, Gurel F, Karaca A. Impact of honeybee and bumblebee pollination on alfalfa seed yield. Acta Agriculturae Scandinavica Section B: Soil and Plant Science. 2008;**58**(1):77-81

[95] TJD W. Heterogeneity among floral visitors leads to discordance between removal and deposition of pollen. Ecology. 1991;**72**:1503-1507

[96] Thomson JD, Goodell K. Pollen removal and deposition by honeybee and bumblebee visitors to apple and almond flowers. Journal of Applied Ecology. 2001;**38**(5):1032-1044

[97] Cane H. Pollinating bees (Hymenoptera: Apiformes) of U.S. alfalfa compared for rates of pod and seed set. Journal of Economic Entomology. 2002;**95**(1):22-27

[98] Batra SW. Comparative efficiency of alfalfa pollination by *Nomia melanderi*, *Megachile rotundata, Anthidium florentinum* and *Pithitis smaragdula* (Hymenoptera: Apoidea). Journal of the Kansas Entomological Society. 1976;**49**: 18-22

[99] Young HJ, Dunning DW, von Hasseln KW. Foraging behavior affects pollen removal and deposition in *Impatiens capensis* (Balsaminaceae). American Journal of Botany. 2007; **94**(7):1267-1271

[100] Brunet J. Holmquist and KGA. The influence of distinct pollinators on female and male reproductive success in the Rocky Mountain columbine. Molecular Ecology. 2009;**18**(17): 3745-3758

[101] Manasse RS. Ecological risks of transgenic plants: Effects of spatial dispersion on gene flow. Ecological Applications. 1992;**2**(4):431-438

[102] Morris WF. Predicting the consequences of plant spacing and biased movement for pollen dispersal by honey bees. Ecology. 1993;**74**(2): 493-500

[103] Ishihama S, Ueno Y, Tsumura WI. Effects of density and floral morph on pollen flow and seed reproduction of an

endangered heterostylous herb, *Primula sieboldii*. Journal of Ecology. 2006;**94**(4): 846-855

[104] Van Deynze AE, Firzpatrick S, Hammon B. Gene Flow in Alfalfa: Biology, Mitigation, and Potential Impact on Production, Council for Agricultural Science and Technology. Iowa, USA: Ames; 2008

[105] Cresswell JE. Estimating the potential for bee-mediated gene flow in genetically modified crop. In: James RR, Pitts-Singer TL, editors. Bee Pollination in Agricultural Ecosystems. New York, NY, USA: Oxford University Press; 2009. pp. 184-202

[106] Cresswell JE, Osborne JL, Bell SA. A model of pollinator-mediated gene flow between plant populations with numerical solutions for bumblebees pollinating oilseed rape. Oikos. 2002; **98**(3):375-384

[107] Johanne B, Christy MS. Impact of bee species and plant density on alfalfa pollination and potential for gene flow. Psyche: A Journal of Entomology. 2010;**12**:145-152. DOI: https://doi.org/10.1155/2010/201858

[108] Fitzpatrick S, Reisen P, Mccaslin M. Alfalfa pollen-mediated gene flow studies, 2000-2001. In: Proc. of the 38th North American Alfalfa Improvement Conference, July 17-21, 2002, Sacromento, California. 2002

[109] Cane JH. A native ground-nesting bee (*Nomia melanderi*) sustainably managed to pollinate alfalfa across an intensively agricultural landscape. Apidologie. 2008;**39**:315-323

[110] Canadian Seed Growers' Association. Regulations and Procedures for Pedigreed Seed Crop Production. Circular 6-94. Section 7. 2003. Available from: www.seedgrowers.ca/regulations

[111] Teuber LRA, Van Deynze S, Mueller. Gene flow in alfalfa under honey bee (*Apis mellifera*) pollination. Invertebrate Zoology. 2004;7:87-93. Available from: https://www.researchgate.net/publication/237561798_Gene_flow_in_alfalfa_under_honey_bee_Apis_mellifera_pollination/link/0deec525ffde8c0874000000/download

[112] Fitzpatrick S, Reisen P, McCaslin M. Pollen-mediated gene flow in Alfalfa: A three-year summary of field research. In: Proceedings of the 2003 Central Alfalfa Improvement Conference, Virtual Meeting July 21-25, 2003. 2003. Available from: http://naaic.org/Meetings/Central2003/Gene_Flow_in_Alfalfa_Abstract_final.doc

[113] Muller MH, Ronfort J, Santoni S, Prosperi JM. Consequence of domestication bottleneck on the DNA sequence diversity of a nuclear gene in wild and cultivated *Medicago sativa*. In: Proceedings of the North American Alfalfa Improvement Conferences (NAAIC). 2002. pp. 55-65

[114] Amand. Risk of alfalfa transgene dissemination and scale dependent effects. TAG. 2000;**101**:107-114

[115] Ontario Ministry of Agriculture, Food and Rural Development. Cutting Management of Alfalfa. 2003. Available from: www.gov.on.ca/OMAFRA/english/crops/field/forages.html

[116] Delaplane K, Mayer DF. Crop Pollination by Bees. Entomologia Experimentalis et Applicata. 2000;**99**:99-103. DOI: 10.1079/9780851994482.0000

[117] Parker FD, Batra SWT, Tepedino VJ. New pollinators for our crops. Agricultural Zoology Reviews. 1987;**2**:279-304

[118] Pitts-Singer TL, Cane JH. The alfalfa leafcutting bee, *Megachile*

rotundata: The world's most intensively managed solitary bee. Annual Review of Entomology. 2011;**56**:221-237. DOI: 10.1146/annurev-ento-120709-14483

[119] Brummer EC. Applying genomics to alfalfa breeding programs. Crop Science. 2004;**446**:1904-1907

[120] Chen F, Dixon RA. Lignin modification improves fermentable sugar yields for biofuel production. Nature Biotechnology. 2007;**257**:759-761

[121] Teotia S, Singh D, Tang X, Tang G. Essential RNA-based technologies and their applications in plant functional genomics. Trends in Biotechnology. 2016;**34**:106-123

Chapter 7

Seed Priming: An Interlinking Technology between Seeds, Seed Germination and Seedling Establishment

Sananda Mondal and Bandana Bose

Abstract

Biologically seed is a small embryonic plant along with either endosperm or cotyledons, enclosed with in an outer protecting covering called seed coat. During the time of seed development large metabolic conversions take place, including proper partitioning of photo-assimilates and the formation of complex polymeric forms of carbohydrate, protein and fats for storing as seed reserves. In developing phase of seeds, every detail information stored in the embryonic plant are genetically and sometimes epigenetically also predetermined and influenced by various environmental/external factors already faced by the mother plant. In the growth cycle of plants, seed germination and seedling establishment are the two critical phases where survivability of the seedlings in natural habitats is a matter of question until the onset of photosynthesis by the established seedling. The various sequence of complex processes known to occur in both the phases i.e., an array of metabolic activities are initiating which eventually leads to the renewal of embryo growth of the dormant seeds and ultimately seedlings are established. Efficient seed germination is an important factor for agricultural sciences and successful establishment of germinated seedling requires a rapid and uniform emergence and root growth. With these aspects of seed physiology kept in mind the present chapter will be designed in such a way where, a gap filling, inter linking, eco- and farmers' friendly technology i.e., 'seed priming' (a pre-sowing partial hydration of seeds) will be considered to improve the rate and uniformity of germination and seedling establishment. Under optimal and adverse environmental conditions, the primed seeds of diversified species lead to an enhanced germination performance with increased vigor index has been reported by various scientists which indicates a good establishment of seedlings in the field and thereafter enhance the performance of crops as a whole.

Keywords: seed, seed development, seed germination, seedling establishment, seed priming

"It's not dead, it's resting."

–Monty Python.

1. Introduction

1.1 Seed and its botany

The seed is a carrier of genetic informations gathered from the past as memory molecules and pass it on future generation of the plants. The natural packaging of all the genetic wisdom within a embryo of any seed is itself a wonder for long periods of time during germplasm collections. In the language of biology, seed is an embryonic plant surrounded by outer protective covering. Embryo is an immature diploid sporophyte developed from a single cell zygote which is covered by nutritive tissue i.e., endosperm in monocots, cotyledon in dicots and in few exceptional cases perisperm which is enveloped by seed coat (testa and tegmen). If we consider the parts of embryo then it consists of an embryonic root called radicle, embryonic shoot known as plumule, contains immature shoot apical meristem, one or more young seed leaves i.e., cotyledons and epicotyl. In a miniature form it also consists a hypocotyl which is a transition region between stem and root [1, 2].

A 'seed' can be defined as any parts of plant if sown in the field should have the potential to regenerate into a new plant [3]. The first critical step of crop production is successful establishment of seedling, and be a determinant factor of the success or failure of the harvested crop. Due to changing climatic conditions, the uncertainty of crop failure is increasing day by day, in this situation to produce quality seeds and maintain the food security of any nation is an issue. For the production of any crop, seed quality is an essential parameter from resource-poor farming to industrial-scale farming. A detailed study is required in terms of sustainability and profitability of crop production with a widely accepted and critically important agronomic trait i.e., seed quality. Most of the crop seeds are tolerant to desiccate condition for examples, orthodox seed are those seeds, can tolerate desiccation and are storable in a dry state for variable periods of time, according to the species. The world-champion orthodox seed is the 2000-year-old Judaean date palm (*Phoenix dactylifera*) seed that was successfully germinated in 2005 [4, 5]. It is an easy process for transportation of seeds in this 'dry' state with a minimal loss in germination potential. The embryo of a seed is a vital pocket where all the genetic information of any crop is stored and act as a delivery system for agricultural biotechnology and crop improvement. For the protection of their investment in crop improvement, seed companies require high quality of seeds to get the maximum benefit while sown in the field. In addition with the purpose of seed companies, the requirement of the farmers to ensure the reliability and successful seedling establishment in their field and get a good return. The readily measureable characteristics of seed quality includes seed viability, purity of the seed lot, health, and mechanical damage, but another essential component is more enigmatic trait of seed vigor as reported by Perry [6].

The present chapter has been prepared on the basis of seed germination and seedling establishment where, an alternative interlinking, gap filling, cost effective, environment- and farmers' friendly technology i.e., 'seed priming' (a pre-sowing partial hydration of seeds) will be considered which has an immense role to improve the speed and uniformity of seed germination and seedling establishment with a good vigor of crop plants.

2. Development of seeds

Development of seeds is an unique attribute for plants reproductive phase, gives them privilege to perpetuate the genetic as well as acquired epigenetic information, generation after generation. Acquiring the knowledge from their mother plant the miniature embryonic plant behaves accordingly and protect themselves against various environmental atrocities, in their whole life span. Physiologically, the development process of a seed is a combined effect of two complex processes i.e., the development of embryo and endosperm is known as embryogenesis. According to Taiz *et al.* [5] the term embryogenesis narrates a process in which a single cell is transformed into a multicellular entity with a characteristic, but typically rudimentary, organization. Within this framework, groups of cells become functionally specialized for the formation of epidermal, cortical, and vascular tissues. Embryogenesis is a linking point between the two phases i.e., gametophytic phase and the differentiation of the sporophytic phase by forming shoot and root meristems. At the initial phase of embryo development, the unicellular zygote elongates and makes a polar structure before dividing and establish a body plan of an embryonic plant along with all acquired information to survive in normal and adverse climatic condition in future and remain as a conserved, arrested phase within the seed for many years. In dicot seed development, majority of the seed volume is occupied by embryo as cotyledons at the time of maturity where the endosperm part is consumed by the embryo during this period of seed development. While studying the monocot seed's structure, it was observed that monocot seeds containing different structural form in comparison to a dicot seed, having a starchy endosperm, occupied most of the spaces inside the seed coat along with an embryo which is positioned at the ventral side of a seed. Moreover, the seed is fully covered with an envelope or seed coat. In addition, seed is an important storage unit and serves nutrition during the time of germination and up to seedling establishment phase as well as supply nutrition to animals and humans in the form of healthy diet [5, 7]. Seed maturation is a key period in a plant's lifecycle allowing for distribution of offspring in space and time. Events occurring during maturation include storage reserve deposition, desiccation, dormancy induction, seed coat formation, and protective compound synthesis. The spatial and temporal regulation of these processes involves the concerted action of a signaling network consisting of hormonal, transcriptional, and metabolic components, which will help in germination process and to concur a good vigorous growth [8].

The seed development process can be divided into four phases where it begins with the fertilization of ovule and ends up with physiological maturity of seeds. In first two phases of seed development viz. Phases I, only cell division and expansion occur (Morphogenesis phase). Accumulation of reserves takes place in Phase II as the dry mass of seeds increases (Maturation phase). In addition, the end phase of seed development is the desiccation phase where seed moisture loss is intensified (Phase III i.e., desiccation phase) [5]. In the reproductive phase of plants, after the fertilization process is over, there is a period to form the structure of seed and as a result cell division, cell expansion and differentiation (histo-differentiation) is observed in which primordia are formed and parts of the embryo can be visualized in future. During this phase, the embryonic cells receive assimilates from the parent plant and as a result there is a significant increase in seed size along with this the moisture content of seed remains constant and high in this phase. At the end of the maturation phase, there is a significant decrease in seed moisture content was observed when cell membrane structure organization is changed and increases in enzyme synthesis was taking place and the seed become metabolically inactive by the changes of plant growth regulator takes places within the cell and they enter dormancy, for successful germination in the next season [2]. Between the embryogenesis and seed germination there is typically a period of seed maturation which is followed by quiescence, where seed dissemination occurs. Germination is delayed until the favorable conditions i.e., water, oxygen, and temperature will arise properly, required for seedling growth. Some special seeds are there where an additional treatment, like light or physical abrasion is a requirement, before their germination.

Whereas, germination comprises all the events which take place between the start of imbibition of the dry seed and the emergence of the embryo, basically the radicle by rupturing the seed coat [5].

3. Seed germination and seedling establishment

Seed germination is a critical phase of plant life cycle as this process initiate a new life [9–11]. This phase of plant's life of any species is a determining factor that germinating new seedling will survive or not in natural habitat. After completion of maturity phase of seed development, seed undergoes in dormancy phase, consists of an embryonic plant and a nutritive storage tissue which can be utilized during the time of germination and seedling establishment till the onset of photosynthesis by the seedling. 'Germination' is a complex process and it is a combination of both physical and biological processes which starts with the imbibition and ends up with the protrusion of radicle where an array of metabolic activities are initiated that leads to the reactivation of the growth of dormant embryo and finally establishes the seedling. During the onset of germination, several sequential events are going on continuously that are triggered by some endogenous factors like the information acquired from mother plant during the time of seed-development process as well as depends on some external environmental factors like water, air, temperature, soil condition etc. For example, *Stellaria* species growing in dry grasslands and in shady deciduous forests are able to germinate some seeds, while those growing in open forest show complete dormancy during maturity in early summer [12]. Similarly, in *Nothofagus* species increasing the altitudinal gradient the seeds that produce may vary considerably in their germination [13]. Study demonstrates no significant result in terms of phylogenetic signal for the germination of *Caragana* species. Seeds of species from arid and semiarid habitats lacked dormancy, seeds of species from semi-humid habitats had physiological dormancy or physical (seed coat) and physiological dormancy, while seeds of species from humid habitats had physical and physiological dormancy. Climate change will definitely affect the germination of seeds, and subsequently recruitment of plants and population dynamics [14].

Germination of seed is an important process for successful seedling establishment, as it is a junction point from where reverse back is not possible. Once the germination has commenced, consumption of accumulated storage reserves was begins which is essential for the seedlings to produce energy and ensure the heterotrophic growth [15–17]. The reserve mobilization phase takes place prior to greening of the cotyledons and as a results depletion of storage reserves, which shift the newly germinating seedling from heterotrophic to autotrophic metabolism, required for successful establishment of seedling [18–19]. Instead of high impact of seedlings performance on crop establishment and seed yield, relatively very little information is available about the molecular processes necessary for the transition phase from seed to seedling, and/or shifting from heterotrophic to (photo) autotrophic growth of seedling. This is a decision-making transition phase for any plants to enter into a natural or agricultural ecosystem which is important for crop production. To reach the phototrophic state, the seed-to-seedling transition stage is regulated by various temporally and spatially employed regulatory factors. These regulatory factors are the controlling unit able to modulate the genes and proteins, may be responsible for seedling establishment. The two distinctive developmental states of germination physiology, i.e., germination and seedling development's gene expression network depicted the global transcriptional interactions. Moreover, this transition stage is characterized with an agronomic trait of seedling establishment and vigor, one of the major aspects in agriculture as revealed by Silva *et al.* [20].

1. In seed germination, uptake of water by seeds is a triphasic process: phase I is characterized with a rapid initial uptake of water i.e., the physical process imbibition. Moreover, Fait *et al.* [15] revealed that in Phase I there is no visible morphological changes was observed, and it is specified as preparation phase for the protrusion of radicle so various kinds of metabolic activities are going on within the seed. Phase I is followed by plateau phase (phase II) where water content is remaining constant but the metabolic activity increases. Protrusion of radicle by rupturing the seed coat, is an embryo-protecting structure acts as a morphological marker to define the end of phase II, and further a sharp increase in water uptake (phase III) was observed as the embryonic axis elongates and the embryo established itself as a young seedling [21].

4. Seed priming as connecting link between seed germination and seedling establishment

Seed priming is an innovative, delicate and complex alternative option for the overall improvement of germination physiology as well as many traits during the period of plants life cycle [22, 23]. Seed priming, pre-activate the physiological metabolic processes of a seed that triggers the germination process by imbibing the seed up to a certain level. In priming, the imbibition should be halted precisely at a right time; this time depends on the species, genotype, and the types of seed and has to be dried carefully under the fan/forced air retaining the original weight. Different kinds of seed priming techniques are applied now a days in various crop plants where various priming agents like osmoregulators, salts, plant growth regulators, biopriming agents, magnetic waves, nanoparticle, macro and micronutrients including water are extensively used according to the specific problem associated with the crop's genetic makeover as well as to survive in the multifarious environmental conditions. With special reference to seed germination, if a graph is plotted between seed water content (imbibition/osmosis) and time among the non-primed and primed seeds in three subsequent phases in a stepwise manner it was observed that the first phase of germination represented by the entry of water in the seed via imbibition; is same for both the cases. Second phase (phase II) represents the hydration process in case of nonprimed seeds. Whereas in case of primed seed, this hydration treatment permits restricted imbibition and induce the pre-germinative metabolism ("activation"), but emergence of radicle is prevented. The last phase (phase III) represents the germination and post-germination processes which resembles each other i.e., in both primed and non-primed seeds [24]. In context to this, Ruttanaruangboworn et al. [25] reported that the pattern of rice seed imbibition was affected by KNO_3 concentration. Higher the concentrations of KNO_3, delayed the time of imbibition of rice seed which took a longer time to reach the end of phases I and II as compared to lower concentration. While using distilled water to see the imbibition pattern, it was quite similar in both rice cultivars. Seeds, primed with 1% KNO_3 during imbibition time of early phase II, improve the seed germination and increased both the speed and uniformity of seed germination.

Seed priming is of different type based on the priming agents used for this purpose, namely hydropriming, osmopriming, halopriming, hormonal priming, nutri-priming, bio-priming, matrix priming, nano-priming, magneto-priming and many other. It can be depicted like hydro-priming (continuous or successive addition of a limited amount of water to the seeds), osmo-conditioning or osmo-priming (exposing seeds to relatively low external water potential), halopriming (pre-sowing soaking of seeds in salt solution), hormonal priming (priming solutions containing limited amount of plant growth regulators or hormones), nutri-priming (seeds are soaked in solutions containing the plant growth-limiting nutrients instead of being soaked just in water), bio-priming (coating of seeds with biocontrol agents), redox priming (it represents the redox state of cell and regulates the key processes in growth and development as well as stress tolerance in response to any external stimuli; plants modify their redox state, and the extent of change is dependent on the nature of the stimulus itself, the dose and the time to which the tissue is exposed), solid matrix priming (mixing seeds with a solid or semisolid material and measured amount of water), and pre-sowing soaking (soaking of seeds either in water or in any solution of low water potential before sowing) [26]. For instance, Bose and Tandon [27] depicted that priming can induce the seed germination by improving the speed and synchronization of germination; it can also increase the seed vigor which requires a very short span or no activation time during germination. A wide range of temperature for germination was experienced by Anaytullah and Bose [28] in wheat, which can help to break the dormancy, or may shorten the emergence time with improved seedling vigor in rice [29] and this leads to a better crop establishment with higher yields in rice [30]. The phase I of priming represented the activation of priming memory molecules, repairing of DNA and mitochondria, increase the respiration, and energy metabolism process, ROS signaling and antioxidant mechanism, transcription and translation of different genes, initiate the cell cycle, and induction of stress response gene like LEA, DHY, AQP, and hormone signaling [31]. In seed priming case, Phase II of germination involved in second rehydration and protein synthesis by using newly synthesized mRNA in phase I. Phase III is denoted as postgermination phase; the events taking place in this phase are stored reserve mobilization, elongation of radicle cells, and ultimately at the end of this phase, radicle is emerging out by rupturing the seed coat [31]. However, Dahal *et al.* [32] observed that if the seeds were pre-treated with different priming agents, it facilitates the absorption process of ionic molecules actively with greater availability of ATP and repairing of deteriorated seed parts was also noted by reducing the leakage of metabolites which leads to faster growth of embryo. Greater cellular membrane integrity along with counteraction of free radical and lipid peroxidation reactions [33], was also reported as a positive effect of seed priming; in addition to this, reactive oxygen species (ROS) chain reaction are mostly found to be directly correlated for the maintenance of viability and reduce the moisture uptake by hydrated-dehydrated seed [34], antipathogenic effects [35], repair of biochemical lesions by the cellular enzymatic repair system [36] and metabolic removal of toxic substances [37], biochemical changes such as activation of enzymes [38], and rate of germination of old seeds was increased [39, 40]. Several scientists depicted that seed priming promotes early replications of DNA, increased the synthesis of RNA and protein and enhanced the growth of embryo [31, 41]. Seed priming is an important viable technology for enhancing uniform and rapid emergence of seedlings and high seedling vigor with a better seed yield in some of the field crops like rice, wheat, maize, chickpea etc. was also reported [42–45].

5. Seed priming and seedling establishment

Successful establishment of seedling is a paramount contributing facet for many developing countries due to low crop yield/production [46]. In most of the developing countries the irrigation facility are not available properly and the crop production is mostly based on rain fed cultivation; as a result water scarcity is there during the time of seed germination and seedling establishment. According to Fischer and Turner [47] high speed and uniform germination of seed under water deficiency is a determinant factor which affect the crop establishment. However, if the effect of stress can be omitted during the time of germination phase, there will be high chances to attain a good crop establishment [48]. Seed priming is an alternative, eco- and farmers' friendly pre-sowing technology promotes the seedling development by modulating and regulating the pre-germination metabolic activity prior to emergence of the radicle and ultimately increases the germination rate and seedlings performance [29].

In continuation with this, *Arabidopsis* is taken as a model plant to study the functional genes involved in seed germination extensively. In this study, *Arabidopsis* mutant was taken into consideration [49–52]. Some of the mutations slow down the germination of seed but do not significantly arrest it. These phenomena may take place due to the mobilization of storage reserves of seeds accumulated during the time of seed maturation and as a result of, a significant reduction in oil reserve content [50, 53, 54]. Whereas, another experiment with a *Arabidopsis* mutant depicted that it was deficient in plastidic pyruvate kinase (pkp1) activity as a result the germination of seeds was delayed and, consequently, it can be said that seedling establishment is not only affected by mobilization of reserves accumulated during seed development, but reduced pyruvate kinase activity was also a determinant factor during germination [55]. In addition, an integration of metabolite and transcript status is required during the time of phase transitions i.e., from seed to seedling. Moreover, Borisjuk *et al.* [56, 57] stated that elevated levels of glucose or sucrose are associated with seedling establishment. For instance, the other metabolic processes of a cell are also working in a high carbohydrate contents state, which may play an important role in seedling establishment [58]. In addition, sugar pathways are also linked with nitrogen pathway via cross-talking mechanism and playing a vital role in seedling establishment [59].

The major transportable form of nitrogen in plants is amino acids [60, 61] and the plant's growth is also dependent on the supply of nitrogen, assimilation and its utilization [62]. According to Zheng, [63] the gene responsible for the synthesis of carbohydrate and proteins are modulated by the balance between C/N and it is a decision-making factor. Likewise, during heterotrophic to autotrophic transition the metabolic regulations extend beyond the primary metabolism. Moreover, the identification of primary metabolites involved in gene expression is possible, by studying their expression pattern of specific genes [64].

For the priming of seeds, mainly in case of halo- and osmo-priming different low water potential chemical solutions are used where PEG, NaCl and various salts can be taken as priming agents. From these solution seeds can uptake various nutrients along with the water during the time of imbibition. These nutrients are required for metabolic activities at the time of germination and seedling establishment in the field and thereafter enhance the performance of crops as a whole [65–67]. These nutrients while going inside the seed may act as a secondary signaling element and enhance the metabolic activity during germination and make it faster as compared to non-primed seeds. Various scientists when compared the non-primed seeds and primed seeds of rice, wheat and mustard, then they observed that seed priming have the capacity to enhance the seedling establishment and seedling vigor, i.e., density of plants, fertile tillers number, test weight, number of grains per panicle, etc. in the field condition [65, 68].

Ghassemi-Golezani *et al.* [69] while studying the early emergence and stand establishment of lentil (*Lens culinaris* Medik) noted that these parameters are playing a vital yield-contributing factors in rainfed areas. To mitigate the problem of rainfed cultivation a seed invigoration technique can be used for lentil seeds; i.e., lentil seed are primed with water (hydropriming) and NaCl (osmo-priming) and resulted in higher seedling emergence and field establishment, as compared to non-primed seeds and primed with PEG and KNO_3. The rate of seedling emergence was also improved in primed seeds with water, NaCl and KNO_3. Therefore, they concluded that hydropriming is a very simple, low cost and environmentally friendly technique for the betterment of seed and seedling vigor of lentil.

Germination and subsequent seedling growth of many springs sown legume varieties, can be inhibited by different genetic and environmental factors depending upon the species and varieties. Under drought condition, priming may be a beneficial technology to reduce the risk of poor stand establishment and it allows more uniform growth under irregular rainfall. Furthermore, pre-hydration of the seeds and enhancing the metabolic activity, hydropriming is a simplistic approach which minimizes the use of chemicals. In this case, the beneficial effects of hydropriming under water stress conditions in lentil were clearly observed [70]. In addition, hydroprimed seeds were germinated and grew more rapidly as compared to non-primed control seeds, and the lentil germinated seeds are benefited by successful seedling establishment and improved seedling growth under water stress condition [70].

Eskandari and Kazemi, [71] did an experiment with cowpea (*Vigna unguiculata*) and considered the parameter early emergence, stand establishment and vigor of seedlings in rainfed areas by priming the seeds with water (hydropriming), 1.5% KNO_3 and 0.8% NaCl (halopriming). It was analyzed that hydroprimed seeds showed significantly improvement in germination rate, seed vigor index, and seedling dry weights followed by NaCl priming. Likewise, it was again proved that in pulses like lentil, cowpea and in many other hydropriming is a better option to avoid chemicals for the improvement seed and seedling vigor which is a simple, cost effective and eco-friendly technology.

For instance, Peraza-Villarreal *et al*. [72] evaluated the effects of seed priming and soil retainers on seed germination and performance of early seedling of useful species in a tropical semideciduous forest of Veracruz, Mexico. They tried to determine the mass, water and lipid content in the seeds of *Albizia saman*, *Cedrela odorata*, *Enterolobium cyclocarpum*, and *Swietenia macrophylla*. The seeds were hydroprimed with water and natural primed by seed burial inside the soil and let them germinate at 25°C and 25/35°C. The seedlings produced were grown in a shade house and planted in a plain terrain land and/or in hillside (slope 75%). After assessing the above said parameters, it was observed that *S. macrophylla* and *E. cyclocarpum* seeds had the lowest and highest water content, whereas, *S. macrophylla* and *C. odorata* had oils in seeds. In addition to this, *A. saman* and *E. cyclocarpum* seeds exhibit physical dormancy. Natural priming improved the rate of germination in *A. saman*, *C. odorata*, and *S. macrophylla,* while hydropriming at 25/35°C enhanced the rate of germination in *E. cyclocarpum*. So natural priming can be recommended for *A. saman*, *C. odorata*, and *S. macrophylla*, and for *E. cyclocarpum* seeds hydropriming was a better option for seed germination and seedling establishment in plain terrain and hillside land. From this study it can be concluded that seed pre-treatments were an inexpensive and easy tool and can be potentially applicable in restoration and conservation programs.

Various works related to seed priming in different crops and in various aspect such as in normal and stressful environmental condition was reported by different scientists. Anaytullah & Bose, [28], Mondal *et al*. [29], Sharma & Bose, [73], Bose *et al*. [74] & Anaytullah *et al*. [75] were working in the field of seed priming with different nitrate salts where the seeds of various fields crops like rice, wheat, mustard were imbibed in nitrate salt for restricted period of time i.e., before the protrusion of radicle & dried back to its original weight under forced air at normal room temperature before sowing. These scientists stated that nitrate seed priming technology have the potentiality to improve the germination physiology, vegetative growth & increase the productivity of these crops as reported by Bose and Mondal, [76]. They named these seed priming process as nitrate seed hardening technology. Bose *et al.* [77] did an experiment in maize (*Zea mays* L.) by soaking the seeds in Ca $(NO_3)_2$ and GA_3 and found a sharp increase in fresh and dry weights of shoot and root during the later period of seedling growth; whereas seed primed with GA_3 showed better performance during early seedling emergence. Likewise, Krishnotar *et al.*, [78] revealed that seed invigoration with $Mg(NO_3)_2$ and distilled water in maize significantly improved the seedling emergence and stand establishment in field condition, vegetative growth, relative water content of shoots and roots, and yield attributes; that means it has an immense crucial role in seedling emergence and stand establishment. In addition, Pant and Bose, [79] reported that seed priming with salts like $NaHCO_3$, KH_2PO_4 and K_2SO_4 in different concentrations, improved the germination percentage, dry weight and vigor index of rice seeds and seedlings respectively. Priming also increased the α-amylase activity and soluble sugar content in the rice endosperm during the time of germination which signifies the importance of seed priming in seed germination to seedling establishment. Moreover, Kumar *et al.* [80] designed an experiment on late sown sesame crop. The sesame seeds were primed with KNO_3 which promotes the germination percentage, shoot and root length, fresh and dry weights of seedling, increase the activity of catalase and peroxidase, proline content as well as different yield attributes. It can be said that seed priming under late sown condition also improves the germination to seedling establishments and as a result yield was increased. For instance, Jangde *et al.* [81] experience low temperature in wheat crop by primed them with $CaCl_2$, $Ca(NO_3)_2$, EGTA during the time of germination and seedling emergence. Among the priming treatments it was noted that $CaCl_2$ was more efficient for the improvement of germination percentage, relative and absolute water content, increased the activity of α-amylase and as a result better shoot and root length of seedlings under low temperature condition. Whereas, Kumar *et al.* [82] depicted the beneficial role of seed priming with water (as hydropriming), $Mg(NO_3)_2$ and $Ca(NO_3)_2$ (as halopriming) for the improvement of seed germination and early seedling establishment of wheat under heavy metal stress (by applying $HgCl_2$) in wheat.

6. Conclusion

To understand the transition between seed to seedling establishment, magnificent number of successful and contributory research works are still going on. Introduction of gene expression and related regulatory networking system provides a template for the identification of different transcriptional regulators beside their cross-linking effects and interacting path ways; it further unveils a no. of signaling mechanisms between germination to seedling formation of a seed. This involves the extensive study and understanding of the physiology of germination. However, agriculturally important but a complex trait which comes after the establishment of seedling i.e., seed vigor, very much responsible, contributory and deciding factor for the productivity of any crop in the farmer's field. From the physiological point of view if we engross ourselves to understand germination basically incepts from the beginning of radicle protrusion and ends to seedling establishment, in that seed vigor has a crucial role, can be practically enriched by using an alternative, interlinking, gap-filling eco- and farmers friendly technology i.e., seed priming. Rapidly growing population with erratic changes in climatic conditions, people are standing in endangered situations; in this point of view the significance of seedling stand establishment and vigor of any crop is increasing with time. The present chapter highlights mainly that how seed priming, a technology, can be handled by general crop growers without much effort is being helpful in reducing the risk of poor seedling establishment in field even in erratic ambient or the broad spectrum of climatic conditions via improving seed vigor. In seed priming, various range of priming agents including water are used to pre-fabricate the seeds. The time duration for priming depends on the genotype makeover of any crop and during this period seed absorb different ions based on the external solution which may act as a signaling element within the seed for the pre-activation of metabolic activities during the process of germination. In nutshell, seed priming interlinks the processes as whole like pre-activation of metabolic events which reduce the time gap from seed sowing to seed germination and speed up the different phases in later one; along with this it also helps a continued supply of mobilized reserve nutrients, required for successful establishment of seedlings.

Conflict of interest

The authors declare no conflict of interest.

Notes/thanks/other declarations

We are thankful to the editor for consideration of our chapter in IntechOpen Book.

Author details

Sananda Mondal[1*] and Bandana Bose[2]

1 Department of Crop Physiology, Institute of Agriculture, Visva-Bharati, Sriniketan, India

2 Department of Plant Physiology, Institute of Agricultural Sciences, Banaras Hindu University, Varanasi, India

*Address all correspondence to: mondalsananda@gmail.com

References

[1] Baskin JM, Baskin CC. Physiology of dormancy and germination in relation to seed bank ecology. In: Leck MA, Parker VT, Simpson RL, editors. Ecology of Seed Banks. New York: Academic press; 1989. pp. 53-69

[2] Bareke T. Biology of seed development and germination physiology. Advances in Plants and Agriculture Research. 2018;**8**(4):336-346. DOI: 10.15406/apar.2018.08.00336

[3] https://en.wikipedia.org/wiki/Seed#cite_ref-2

[4] Gray DJ. Seed Development. Encyclopedia of Applied Plant Sciences. 2003:1305-1310. DOI: 10.1016/B0-12-227050-9/00034-X

[5] Taiz L, Zeiger E, Moller IM, Murphy A. Plant Physiology and Development. Sixth ed. Inc. USA: Sinauer Associates; 2015

[6] Perry DA.1980. The concept of seed vigour and its relevance to seed production techniques. In: Hebblethwaite PD, editor. Seed production. London: Butterworths; 1980. p. 585-591

[7] Mahto A, Mathew IE, and Agarwal P. Decoding the transcriptome of rice seed during development. 2017. DOI:10.5772/intechopen.70574

[8] Sabelli PA. Seed development: A comparative overview on biology of morphology, physiology and biochemistry between monocot and dicot plants. In: Seed development: OMICS technologies toward improvement of seed quality and crop yield. Springer; 2012. pp. 3-25

[9] Liu Y et al. Nitric oxide-induced rapid decrease of abscisic acid concentration is required in breaking seed dormancy in Arabidopsis. New Phytologist. 2009;**183**:1030-1042. DOI: 10.1111/j.1469-8137.2009.02899.x

[10] Norden N et al. The relationship between seed mass and mean time to germination for 1037 tree species across five tropical forests. Functional Ecology. 2009;**23**:203-210. DOI: 10.1111/j.1365-2435.2008.01477.x

[11] El-Maarouf-Bouteau H et al. Reactive oxygen species, abscisic acid and ethylene interact to regulate sunflower seed germination. Plant, Cell & Environment. 2015;**38**:364-374. DOI: 10.1111/pce.12371

[12] Vandelook F, Van de Moer D, Van Assche JA. Environmental signals for seed germination reflect habitat adaptations in four temperate Caryophyllaceae. Functional Ecology. 2008;**22**:470-478. DOI: 10.1111/j.1365-2435.2008.01385.x

[13] Arana MV et al. Seed dormancy responses to temperature relate to *Nothofagus* species distribution and determine temporal patterns of germination across altitudes in Patagonia. New Phytologist. 2016;**209**:507-520. DOI: 10.1111/nph.13606

[14] Fang X-W et al. Seed germination of Caragana species from different regions is strongly driven by environmental cues and not phylogenetic signals. Scientific Reports. 2017;7:11248. DOI: 10.1038/s41598-017-11294-x

[15] Fait A, Angelovici R, Less H, Ohad I, Urbanczyk-Wochniak E, Fernie AR, et al. Arabidopsis seed development and germination is associated with temporally distinct metabolic switches. Plant Physiology. 2006;**142**:839-854. DOI: 10.1104/pp.106.086694

[16] Carrera E, Holman T, Medhurst A, Peer W, Schmuths H, Footitt S, et al.

Gene expression profiling reveals defined functions of the ATP-binding cassette transporter COMATOSE late in phase II of germination. Plant Physiology. 2007;**143**:1669-1679. DOI: 10.1104/pp.107.096057

[17] Bassel GW, Fung P, Chow TF, Foong JA, Provart NJ, Cutler SR. Elucidating the germination transcriptional program using small molecules. Plant Physiology. 2008;**147**:143-155. DOI: 10.1104/pp.107.110841

[18] Mansfield SG, Briarty LG. The dynamics of seedling and cotyledon cell development in Arabidopsis thaliana during reserve mobilization. International Journal of Plant Sciences. 1996;**157**:280-295. DOI: 10.1086/297347

[19] Allen E, Moing A, Ebbels TM, Maucourt M, Tomos AD, Rolin D, et al. Correlation network analysis reveals a sequential reorganization of metabolic and transcriptional states during germination and gene-metabolite relationships in developing seedlings of Arabidopsis. BMC Systems Biology. 2010;**4**:62. DOI: 10.1186/1752-0509-4-62

[20] Silva AT, Ribone PA, Chan RL, Ligterink W, Hilhorst HWM. A predictive coexpression network identifies novel genes controlling the seed-to-seedling phase transition in *Arabidopsis thaliana*. Plant Physiology. 2016;**170**:1-14. DOI: 10.1104/pp.15.01704

[21] Bewley JD, Bradford KJ, Hilhorst HWM, Nonogaki H. Seeds: physiology of development, germination and dormancy. 3rd ed. New York: Springer; 2013. DOI: 10.1007/978-1-4614-4693-4

[22] Mondal S, Bose B. Impact of micronutrient seed priming on germination, growth, development, nutritional status and yield aspects of plants. Journal of Plant Nutrition. 2019;**42**:2577-2599. DOI: 10.1080/01904167.2019.1655032

[23] Waqas M et al. Advances in the concept ans methods of seed priming. In: Priming and pretreatment of seeds and seedlings. Singapore: Springer Nature, Pvt. Ltd;

[24] Rajjou L, Duval M, Gallardo K, Catusse J, Bally J, Job C, et al. Seed germination and vigor. Annual Review of Plant Biology. 2012;**63**:507-533. DOI: 10.1146/annurev-arplant-042811-105550

[25] Ruttanaruangboworn A, Chanprasert W, Tobunluepop P, Onarmol D. Effect of seed priming with different concentrations of potassium nitrate on the pattern of seed imbibition and germination of rice (*Oryza sativa* L.). Journal of Integrative Agriculture. 2017;**16**(3):605-613. DOI: 10.1016/S2095-3119(16)61441-7

[26] Bose B, Kumar M, Singhal RK, Mondal S. In: Rakshit A, Singh HB, editors. Advances in Seed Priming. Singapore: Springer Nature; 2018. DOI: 10.1007/978-981-13-0032-5

[27] Bose B, Tandon A. Effect of magnesium nitrate on metabolism in germinating maize seeds. Indian Journal of Plant Physiology. 1991;**34**:69-71

[28] Anaytullah BB. Nitrate-hardened seeds increase germination, amylase activity and proline content in wheat seedlings at low temperature. Physiology and Molecular Biology of Plants. 2007;**13**(3&4):199-207

[29] Mondal S, Vijai P, Bose B. Role of seed hardening in rice variety Swarna (MTU 7029). Research Journal of Seed Science. 2011;**4**:157-165. DOI: 10.3923/riss.2011.157.165

[30] Srivastava AK, Bose B. Effect of nitrate seed priming on phenology, crop growth rate and yield attributes in rice

(*Oryza sativa* L.). Vegetos. 2012;**25**: 174-181

[31] Chen K, Arora R. Priming memory invokes seed stress-tolerance. Environmental and Experimental Botany. 2013;**94**:33-45. DOI: 10.1016/j.envexpbot.2012.03.005

[32] Dahal P, Bradford KJ, Jones RA. Effects of priming and endosperm integrity on seed germination rates of tomato genotypes II. Journal of Experimental Botany. 1990;**41**:1441-1453. DOI: 10.1093/jxb/41.11.1431

[33] Rudrapal AB, Basu RN. Use of chlorine and bromine in controlling mustard seed deterioration. Seed Research. 1982;**9**:188-191

[34] Dollypan BRN. Mid-storage and pre-sowing seed treatments for lettuce and carrot. Scientia Horticulturae. 1985;**33**:1026-1027. DOI: 10.1016/0304-4238(85)90072-X

[35] Powell AA, Mathews S. Cell membranes and seed leachate conductivity in relation to the quality of seeds for sowing. Journal of Seed Technology. 1986;**10**:81-100

[36] Villiers TA, Edgcumbe DJ. On the cause of seed determination in dry storage. Seed Science and Technology. 1975;**3**:761-774

[37] Basu RN, Chattopadhyay K, Pal P. Maintenance of seed viability in rice (*Oryza sativa* L.) and Jute (*Corchorus capsularis* L. and *C. olitorius* L.). Indian Agriculture. 1973;**18**:76-79

[38] Mondal S, Bose B. Kinetics studies on α-amylase extracted from germinating wheat endosperm of primed and non-primed seeds. Indian Journal of Agricultural Biochemistry. 2012;**25**:137-141

[39] Gray D, Steckel JR. Seed quality in carrots: the effects of seed crop plant density, harvest date and seed grading on seed and seedling variability. Journal of Horticultural Sciences. 1983;**58**: 393-401

[40] Lee SS, Kim JH, Hong SB, Yun SH, Park EH. Priming effect of rice seeds on seedling establishment under adverse soil conditions. Korean Journal of Crop Science. 1998;**43**:194-198

[41] Gianella M et al. Molecular aspects of seed priming as a means of progress in crop improvement. In: Advancement in Crop Improvement Techniques. ScienceDirect; 2020. pp. 89-100. DOI: 10.1016/B978-0-12-818581-0.00006-1

[42] Basra SMA, Zia MN, Mehmood T, Afzal I, Khaliq A. Comparison of different invigoration techniques in wheat seeds. Pakistan Journal of Arid Agriculture. 2002;**5**:11-16

[43] Chiu KY, Chen CL, Sung JM. Effect of priming temperature on storability of primed sh-2 sweet corn seed. Crop Science. 2002;**42**:1996-2003. DOI: 10.2135/cropsci2002.1996

[44] Harris D, Joshi A, Khan PA, Gothkar P, Sodhi PS. On farm seed priming in semi-arid agriculture development and evaluation in maize, rice and chickpea in India using participatory methods. Experimental Agriculture. 1999;**35**:15-29. DOI: 10.1017/S0014479799001027

[45] Murungu FS, Chiduza C, Nyamugafata P, Clark LJ, Whalley WR, Finch-Savage WE. Effects of 'on-farm seed priming' on consecutive daily sowing occasions on the emergence and growth of maize in semi-arid Zimbabwe. Field Crops Research. 2004;**89**:49-57. DOI: 10.1016/j.fcr.2004.01.020

[46] Hadas A. Germination and seedling establishment. Bet Dagan, Israel: The Volcani Center, Elsevier Ltd; 2005. p. 130

[47] Fischer RA, Turner NC. Plant productivity in the arid and semiarid zones. Annual Review of Plant Physiology. 1978;**29**:277-317. DOI: 10.1146/annurev.pp.29.060178.001425

[48] Ashraf M, Rauf H. Inducing salt tolerance in maize (*Zea mays* L.) through seed priming with chloride salts, growth and ion transport at early growth stages. Acta Physiologiae Plantarum. 2001;**23**:407-414. DOI: 10.1007/s11738-001-0050-9

[49] Debeaujon I, Koornneef M. Gibberellin requirement for Arabidopsis seed germination is determined both by testa characteristics and embryonic abscisic acid. Plant Physiology. 2000;**122**:415-424. DOI: 10.1104/pp.122.2.415

[50] Lu C, Hills MJ. Arabidopsis mutants deficient in diacylglycerol acyltransferase display increased sensitivity to abscisic acid, sugars, and osmotic stress during germination and seedling development. Plant Physiology. 2002;**129**:1352-1358. DOI: 10.1105/tpc.019646

[51] Fulda M, Schnurr J, Abbadi A, Heinz E, Browse J. Peroxisomal Acyl-CoA synthetase activity is essential for seedling development in Arabidopsis thaliana. Plant Cell. 2004;**16**:394-405. DOI: 10.1105/tpc.019646

[52] Yang M, Zhang B, Jia J, Yan C, Habaike A, Han Y. RRP41L, a putative core subunit of the exosome, plays an important role in seed germination and early seedling growth in Arabidopsis. Plant Physiology. 2013;**161**:165-178. DOI: 10.3389/fpls.2017.00981

[53] Penfield S, Graham S, Graham IA. Storage reserve mobilization in germinating oilseeds: arabidopsis as a model system. Biochemical Society Transactions. 2005;**33**:380-383. DOI: 10.1042/BST0330380

[54] Focks N, Benning C. wrinkled1: A novel, low-seed-oil mutant of Arabidopsis with a deficiency in the seed-specific regulation of carbohydrate metabolism. Plant Physiology. 1998;**118**:91-101. DOI: 10.1104/pp.118.1.91

[55] Andre C, Benning C. Arabidopsis seedlings deficient in a plastidic pyruvate kinase are unable to utilize seed storage compounds for germination and establishment. Plant Physiology. 2007;**145**:1670-1680. DOI: 10.1104/pp.107.108340

[56] Borisjuk L, Walenta S, Rolletschek H, Mueller-Klieser W, Wobus U, Weber H. Spatial analysis of plant metabolism: sucrose imaging within *Vicia faba* cotyledons reveals specific developmental patterns. Plant Journal. 2002;**29**:521-530. DOI: 10.1046/j.1365-313x.2002.01222.x

[57] Borisjuk L, Walenta S, Weber H, Mueller-Klieser W, Wobus U. High-resolution histographical mapping of glucose concentrations in developing cotyledons of *Vicia faba* in relation to mitotic activity and storage processes: glucose as a possible developmental trigger. Plant Journal. 1998;**15**:583-591. DOI: 10.1104/pp.107.108340

[58] Tognetti JA, Pontis HG, Martínez-Noël GMA. Sucrose signaling in plants: a world yet to be explored. Plant Signaling Behaviour. 2013;**8**(3):e23316. DOI: 10.4161/psb.23316

[59] Coruzzi GM, Zhou L. Carbon and nitrogen sensing and signalling in plants: emerging 'matrix effects'. Current Opinion in Plant Biology. 2001;**4**:247-253. DOI: 10.1016/S1369-5266(00)00168-0, 10.1016/s1369-5266(00)00168-0

[60] Tegeder M, Ward JM. Molecular evolution of plant AAP and LHT amino acid transporters. Frontiers in Plant

Science. 2012. DOI: 10.3389/fpls.2012.00021

[61] Tegeder M. Transporters involved in source to sink partitioning of amino acids and ureides: opportunities for crop improvement. Journal of Experimental Botany. 2014;**65**:1865-1878. DOI: 10.1093/jxb/eru012

[62] Stitt M, Krapp A. The interaction between elevated carbon dioxide and nitrogen nutrition: the physiological and molecular background. Plant, Cell & Environment. 1999;**22**:583-621. DOI: 10.1046/j.1365-3040.1999.00386.x

[63] Zheng ZL. Carbon and nitrogen nutrient balance signaling in plants. Plant Signaling Behaviour. 2009;**4**:584-591. DOI: 10.4161/psb.4.7.8540

[64] Gibon Y, Usadel B, Blaesing OE, Kamlage B, Hoehne M, Trethewey R, et al. Integration of metabolite with transcript and enzyme activity profiling during diurnal cycles in Arabidopsis rosettes. Genome Biology. 2006;7:R76-R76. DOI: 10.1186/gb-2006-7-8-r76

[65] Du LV, Tuong TP. Enhancing the performance of dry-seeded rice: effects of seed priming, seedling rate, and time of seedling. In: Direct seeding: research strategies and opportunities. Manila: International Research Institute; 2002. pp. 241-256

[66] Pandey D, Bose B. Influence of pre-sowing seed treatment with nitrate salts and different sowing dates on performance of mustard. Indian Journal of Plant Physiology. 2006;**11**:261-265

[67] Sharma MK, Bose B, Shrivastava AK. Effect of seed hardening with nitrate salts on physiological attributes at ear head emergence stage and yield of wheat (*Triticum aestivum* L.). International Journal of Agricultural Science. 2009;**5**:439-442

[68] Bose B, Mishra T. Response of wheat seed to pre sowing seed treatment with $Mg(NO_3)_2$. Annals of Agricultural Research. 1992;**13**:132-136

[69] Ghassemi-Golezani K, Asghar Aliloo A, Valizadeh M, Moghaddm M. Effects of different priming techniques on seed invigoration and seedling establishment of lentil (*Lens culinaris* Medik). Journal of Food, Agriculture and Environment. 2008;**6**(2):222-226

[70] Saglam S, Day S, Kaya G, Gurbuz A. Hydropriming increases germination of lentil (*Lens culinaris* Medik.) under water stress. Notulae Scientia Biologicae. 2010;**2**(2):103-106. DOI: 10.15835/nsb224602

[71] Eskandari H, Kazemi K. Effect of Seed Priming on Germination Properties and Seedling Establishment of Cowpea (*Vigna sinensis*). Notulae Scientia Biologicae. 2011;**3**(4):113-116. DOI: 10.15835/nsb346338

[72] Peraza-Villarreal H, Sanchez-Coronado ME, Lindig-Cisneros R, Tinoco-Ojanguren C, Velazques-Rosas N, Camara-Cabrales L, et al. Seed Priming Effects on Germination and Seedling Establishment of Useful Tropical Trees for Ecological Restoration. Tropical Conservation Science. 2018;**11**:1-15. DOI: 10.1177/1940082918817886

[73] Sharma MK, Bose B. Effect of seed hardening with nitrate salts on seedling emergence, plant growth and nitrate assimilation of wheat. Physiology and Molecular Biology of Plants. 2006;**12**(2):173-176

[74] Bose B, Kumar R, Kuril SK, Srivastava AK. Hardening of mustard seeds with magnesium nitrate increase seed germination, vegetative growth, nitrogen assimilation and yield. Brassica. 2007;**9**(1-4):33-38

[75] Anaytullah SAK, Bose B. Impact of seed hardening treatment with nitrate salts on nitrogen & antioxidant defence metabolism in wheat under different sowing conditions. Vegetos. 2012;**25**(1):292-299

[76] Bose B, Mondal S. Climate change and sustainable agriculture in context to seed priming and role of nitrate. Vegetos. 2013;**26**(special):192-204. DOI: 10.5958/j.2229-4473.26.2s.140

[77] Bose B, Tyagi B, Devi TM. Influence of seed soaking with Ca(NO_2) and GA_3 on the growth of seedlings of *Zea mays*. L. (var TuiPakhia). Indian Biologist. 1992;**XXIV**(1)

[78] Krishnotar BB, Srivastava AK, Shah JP. Response of rabi maize crop to seed invigoration with magnesium nitrate and distilled water. Indian Journal of Plant Physiology. 2009;**14**(1):71-77

[79] Pant B, Bose B. Germination of rice seeds primed with various salts. Indian Biologist. 2015;**47**(1):17-21

[80] Kumar S, Hemantranjan A, Mondal S, Bose B. Impact of KNO_3 primed seeds on the performance of late sown sesame. International Journal of Bio-resource and Stress Management. 2016;**7**(4):950-954. DOI: 10.23910/IJBSM/2016.7.4.1418a

[81] Jangde S, Vijay P, Bose B. Influence of calcium salts treatment on growth and germination of wheat during low temperature stress. Phytomorphology. 2014;**64**(1&2):5-14

[82] Kumar M, Pant B, Mondal S, Bose B. Hydro and halo priming: influenced germination in responses in wheat var HUW-468 under heavy metal stress. Acta Physiologiae Plantarum. 2016;**38**:217. DOI: 10.1007/S11738-016-2226-3